THE ECONOMICS OF SCIENCE AND TECHNOLOGY

THE ECONOMICS OF SCIENCE AND TECHNOLOGY:

AN OVERVIEW OF INITIATIVES TO FOSTER INNOVATION, ENTREPRENEURSHIP, AND ECONOMIC GROWTH

Maryann P. Feldman
Johns Hopkins University

Albert N. Link
University of North Carolina at Greensboro

Donald S. Siegel
University of Nottingham, U.K.

Springer Science+Business Media, LLC

Library of Congress Cataloging-in-Publication Data

Feldman, Maryann P.
The economics of science and technology : an overview of initiatives to foster innovation, entrepreneurship, and economic growth / Maryann P. Feldman, Albert N. Link, Donald S. Siegel.
p. cm.
Includes bibliographical references and index.
ISBN 978-1-4613-5335-5 ISBN 978-1-4615-0981-3 (eBook)
DOI 10.1007/978-1-4615-0981-3
1. Research—Economic aspects--United States. 2. Technology—Economic aspects—United States. 3. Science and state—United States. 4. Technology and state—United States. I. Link, Albert N. II. Siegel, Donald S., 1959- III. Title.

Q180.55.E25 F45 2002
338.973'06—dc21 **2002019101**

Originally published by Kluwer Academic Publishers in 2002
Softcover reprint of the hardcover 1st edition 2002

Printed on acid-free paper.

For our families

CONTENTS

LIST OF TABLES

LIST OF FIGURES

ACKNOWLEDGEMENTS

Aspects of the research that underlies this book first appeared as a centerpiece article in the *Journal of Technology Transfer*. That abbreviated version of the material presented herein was prepared as a primer or general introduction to the field that we call the economics of science and technology. A number of eminent scholars were invited to participate in the *JTT* project, and the material contained in this book, although expanded in scope, has greatly benefited from their input and thought. We are appreciative to each of them: David Audretsch of Indiana University, Barry Bozeman of the Georgia Institute of Technology, Kathy Combs of the University of St. Thomas, Paula Stephan of Georgia State University, Greg Tassey of the National Institute of Standards and Technology, and Chuck Wessner of the National Academy of Sciences. We also thank John Jankowski of the National Science Foundation and John T. Scott of Dartmouth College for their comments and suggestions on earlier versions.

And finally, we thank our families, whose patience has been greatly appreciated throughout this entire undertaking and to whom this book is dedicated.

1 INTRODUCTION

Science and technology have long been regarded as important determinants of economic growth. Edwin Mansfield (1971, pp. 1-2), a pioneer in the economics of technological change, noted:

> Technological change is an important, if not the most important, factor responsible for economic growth ... without question, [it] is one of the most important determinants of the shape and evolution of the American economy.

Science and technology are even more important in the "new economy," with its greater emphasis on the role of intellectual property and knowledge transfer. Therefore, it is unfortunate that most individuals rarely have the opportunity to explore the economic implications of science and technology. As a result, the antecedents and consequences of technological change are poorly understood by many in the general public.

This lack of understanding is reflected in a recent survey conducted by the National Science Board (2000), summarized in *Science & Engineering Indicators*.[1] As shown in Table 1.1, the findings of the survey indicated that many Americans, despite a high level of interests in such matters, are not as well-informed about technological issues as they are about other policy issues. As shown in the table, individuals self assess, based on a scale from 1 to 100, their interest in science and technology policy issues as being relatively high, yet they self assess their knowledge or informedness about these issues relatively lower. However, as the quoted passage from Mansfield above indicates, technological change, which does

[1] *Science & Engineering Indicators* is published bi-annually in accordance with its responsibilities as outlined in the National Science Foundation Act of 1950.

follow from science and technology policy, is the most important factor responsible for economic growth. On the other hand, Americans appear to be extremely interested in learning more about science and technology issues.

Table 1.1. Indices of Public Interest and Self-Assessed Knowledge in Selected Policy Issues, 1999

Interest	Policy Issues	Informedness
82	New medical discoveries	53
71	Environmental pollution	48
71	Local school issues	58
67	Issues about new scientific discoveries	44
65	Use of new inventions and technologies	43
65	Economic issues and business conditions	50
64	Military and defense policy	44
55	Use of nuclear energy to generate electricity	29
53	International and foreign policy	40
51	Space exploration	37
47	Agricultural and farm issues	33

Source: National Science Board (2000, Appendix Tables 8-2, 8-5).
Note: Indices are based on a scale of 1 to 100, where 100 denotes the maximum level of interest or of informedness.

The purpose of this book is to fill this gap by providing an overview of key topics in the literature on the economics of science and technology. Considerable attention is devoted to historical and institutional information concerning these topics, almost exclusively from a U.S. perspective. We believe that an understanding of the current science and technology landscape depends to a large extent on an understanding of how the institutions that support science and technology evolved. We also devote attention to the academic literature related to the topics discussed. We do not stress the theoretical foundations of this literature. Instead, we focus on the conclusions of the literature, in an effort to provide the reader with an overview of current thought.

Our goal is for this book to become an initial reference for students interested in the economics of science and technology, as well as for policy makers who find such information fundamental for their involvement in innovation and growth policies.

As a starting point toward this goal, we provide some fundamental definitions of important concepts.

FUNDAMANTAL DEFINITIONS

As with any discipline—and we view the economics of science and technology as a growing field of inquiry that draws on concepts from numerous disciplines—there are a number of fundamental concepts to set forth.

In conversation, terms like *science* and *technology* as well as *invention* and *innovation*, among others, are often used interchangeably. However, for academics and policymakers there are important distinctions between these often-paired terms, and these distinctions give each term a unique meaning.

Science, in a broad sense, is the search for knowledge, and that search is based on observed facts and truths. Thus, science begins with known starting conditions and searches for unknown end results (Nightingale 1998). Technology is the application of new knowledge, learned through science, to some practical problem. Technological change is the rate at which new knowledge is diffused and put into use in the economy.

Closely related to science and technology are the concepts of invention and innovation. Following Bozeman and Link (1983, p. 4):

> The concepts commonly used in connection with innovation are deceptively simple. *Invention* is the creation of something new. An invention becomes an *innovation* when it is put in use.

Innovations may be new products, new processes, or new organizational methods that are novel and add value to economic activity. Thus, at a general level, invention parallels the concept of science and innovation parallels the concept of technology.

It is useful to think of an innovation as something new that has been brought into use. Thus, the innovation represents, in a sense, a new underlying technology. When the innovation is itself the final marketable result, it is sometimes referred to as a product innovation. When the innovation is applied in a subsequent production process, it is sometimes referred to as a process innovation (meaning that its application affects a production process).

Embedded in this distinction between invention and innovation is a process whereby inventions become applied. This process is central to what we call entrepreneurship. Entrepreneurship is a process involving the organization of resources, and the output of that process is an innovation. This process is also referred to as the innovation process with an innovation being the defined output. Of course, for entrepreneurship, or entrepreneurial activity, to have or lead to economic value the resultant output or innovation must itself have economic value.

Figure 1.1 provides an initial look at what we call the entrepreneurial process. We will expand this analysis in subsequent chapters through the inclusion of additional technology elements in the entrepreneurial process. But, the process, as simply illustrated in Figure 1.1 serves here the purpose of introducing systematically several important concepts:

- The entrepreneurial or innovative organization, typically a private firm, has a focus that results from its agreed upon strategic direction. This strategic direction, coupled with competitive market conditions, generates an entrepreneurial response.
- The purposive activity associated with the entrepreneurial response leads to an innovation.
- There are market forces at work that are, in part, beyond the influence of the firm. These forces determine the economic value of the innovation and hence, the value added to the firm as well as to the user of the innovation.

Figure 1.1. The Entrepreneurial Process: An Initial Look

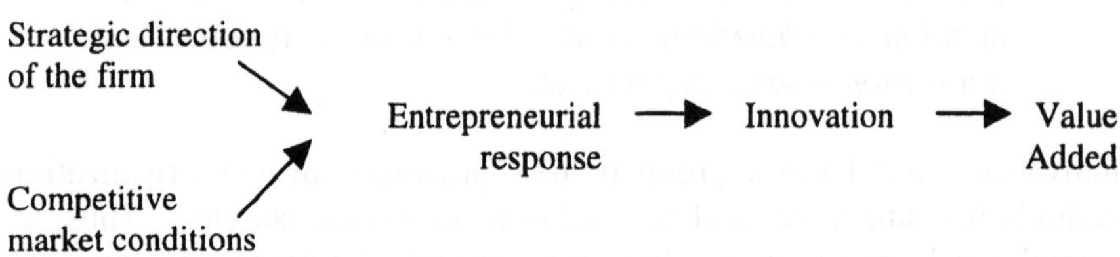

There is a subtle distinction between entrepreneurship, or the innovation process, and the process of science and discovery. As noted above, science moves from starting conditions toward unknown results whereas the innovation process starts with an anticipated intended result and moves toward the unknown starting conditions that will produce it. In a broad sense, knowing or discovering starting conditions is tantamount to identifying innovation policy levers.

OVERVIEW OF THE BOOK

The remainder of the book is organized as follows. Chapter 2 summarizes major public policy initiatives toward science and technology in the United States from the colonial period to the present. Our up-front emphasis on policy underscores the subtlety of partnerships involving the public and private sectors that have emerged over the last three centuries. These so-called public/private partnerships have evolved over time from the government's desire to steer private investment towards certain types of scientific activity and to stimulate the development and use of new technologies throughout the economy. Thus, the federal government has attempted to establish an environment that is conducive for private sector investment in research and development (R&D), as well as one in which the public and private sectors can be partners in undertaking innovative activity.

Chapter 3 emphasizes the role of technology in economic growth and sets the stage for understanding the scope of science and technology policy mechanisms used to maintain national growth. Fundamental to all such policy mechanisms is the relationship between R&D investments and technological advancements, and between technological advancements and economic growth. Dimensions of R&D are described in Chapter 4.

Chapter 5, building on the concepts in this introductory chapter, emphasizes the entrepreneurial nature of firms. That nature is one to innovate as well as to respond to policy initiatives.

Chapter 6 advances an economic rationale for government's role in the innovation process. An articulation of such a role has long been absent from science and technology policy debates.

Having set forth this rationale, four policy mechanisms are discussed from an historical as well as an economic perspective: patents in Chapter 7, tax incentives in Chapter 8, research collaborations in Chapter 9, and public/private partnerships that subsidize research in Chapter 10 and Chapter 11. These mechanisms, and their relationship to firm behavior, are then summarized in Chapter 12.

All of the policy mechanisms discussed in Chapters 7 through 11 are designed to stimulate the private sector's demand for R&D resources. For demand mechanisms to be effective there must also be a supply response in terms of the education and increased availability of scientists and engineers. This is the topic of Chapter 13.

As with all public-sector initiatives, there is also an issue of public accountability. How effectively are public research funds being spent? The history of public accountability and related evaluation methods are described in Chapter 14.

Chapter 15 contains our summary statement regarding the current state of the field of economics of science and technology and future directions for both policy and academic research.

2 AN HISTORICAL PERSPECTIVE

The development of science, technology, and economic growth in the United States was greatly influenced by the scientific discoveries and university infrastructure within Europe during the colonial period. While it is difficult to pinpoint how or which specific elements of scientific and technical knowledge diffused across the Atlantic, certain milestone events can be dated and key individuals can be identified.

This background, as described in Unesco (1968) and National Science Board (2000), from which much of this chapter draws, gives an appreciation for the role that science and technology resources have played in the developing American nation and contributed to shaping the preeminent role that the former colony achieved.

An understanding of these historical events and players is important because it allows one to understand the environment in which innovation takes place and the genesis of the assumptions that underlie the public policy initiatives that influence this environment. It also illustrates the evolving role that the government has played in promoting science and technology, and the resulting entrepreneurial process.

THE COLONIAL PERIOD

The first member of the Royal Society of London to immigrate to the Massachusetts Bay Colony was John Winthrop, Jr. in 1631, just a few years after the founding of the Colony. As a scientist, he is credited with establishing druggist shops and chemistry laboratories in the surrounding villages to meet the demand for medicine.

According to Unesco (1968, p. 9), these ventures were "perhaps the first science based commercial enterprise of the New World."

Before the turn of the eighteenth century, colonists made noticeable advances toward what may be called a scientific society, organizing scientists who came from England and other European countries into communities that promoted scientific inquiry. In 1683, the Boston Philosophical Society was formed to advance knowledge in philosophy and natural history.

Benjamin Franklin formed the American Philosophical Society of Philadelphia in 1742 for the purpose of encouraging correspondence with colonists in all areas of science. This Society later merged with the Franklin-created American Society to promote what Franklin called "useful knowledge," and it still exists today. The combined Society focused on making available advancements in agriculture and medicine to all individuals by sponsoring the first medical school in America (also supported by the Pennsylvania House of Representatives). Thus, Franklin's combined Society was a hallmark of how public and private sector interests could work together for the common weal.

Influenced by the actions of Pennsylvania and later Massachusetts with regard to sponsorship of scientific institutions, the establishment of national universities for the promotion of science was first discussed at the Constitutional Convention in 1787. However, at that time the founders of the Constitution believed educational and scientific activities should be independent of direct national governmental control. But, they felt that the national government should remain an influential force exerting its influence through indirect rather than direct means.

For example, Article I, Section 8, of the Constitution states:

> The Congress shall have the power ... To promote the progress of science and useful arts, by securing for limited times to authors and inventors the exclusive right to their respective writings and discoveries.

Soon thereafter, in 1790, Congress passed the first patent act.

However, Thomas Jefferson championed a more direct role for the government in the area of science. While president, Jefferson sponsored the Lewis and Clark expedition in 1803 to advance the geographic knowledge of the nation, thus making clear that "the promotion of the general welfare depended heavily upon advances in scientific knowledge" (Unesco 1968, p. 11). In fact, this action by Jefferson set several important precedents including the provision of federal funds to individuals for scientific endeavors.

Although the Constitution did not set forth mechanisms for establishing national academic institutions, based on the founders' belief that the government should have only an indirect influence on science and technical advancement, the need for a national institution related to science and technology was recognized soon after the Revolutionary War. For example, West Point was founded in 1802 as the first national institution of a scientific and technical nature, although Connecticut established the first State Academy of Arts and Sciences in 1799.

In the early 1800s, universities began to emphasize science and technical studies, and in 1824 Rensselaer Polytechnic Institute was founded in New York State to emphasize the application of science and technology.

The *American Journal of Science* was the first American scientific publication, followed in 1826 by the *American Mechanics Magazine*.

The social importance of the government having a direct role in the creation and application of technical knowledge was emphatically demonstrated in the 1820s and 1830s through its support of efforts to control the cholera epidemic of 1822. Also during that time period federal initiatives were directed toward manufacturing and transportation. In fact, the Secretary of the Treasury—the Department of the Treasury being the most structured executive department at that time—directly funded the Franklin Institute in Philadelphia to investigate the causes of these problems. This action, driven by public concern as well as the need to develop new technical knowledge, was the first instance of the government sponsoring research in a private-sector organization.

In 1838, the federal government again took a lead in the sponsorship of a technological innovation that had public benefits. After Samuel Morse demonstrated the feasibility of the electric telegraph, Congress provided him with $30,000 to build an experimental line between Baltimore, Maryland, and Washington, DC. This venture was the first instance of governmental support to a private researcher.

In retrospect, one could make an argument that Jefferson's funding of Lewis and Clark was the first instance of public support for pure research, whereas Morse was funded to conduct applied research. Although not discussed herein, there are other historical examples of governmental support to individuals for research that had the potential to benefit society, such as the Longitude Act of 1714. The British Parliament offered a prize (equal to several million dollars in today's terms) for a practicable solution for sailing vessels to determine longitude (Sobel 1995).

Public/private research relationships continued to evolve in frequency and in scope. In 1829, James Smithson, gifted $500,000 to the United States to found an institution in Washington, DC for the purpose of "increasing and diffusing knowledge among men" (Unesco, p. 12). Using the Smithson gift as seed funding, Congress chartered the Smithsonian

Institution in 1846, and Joseph Henry became its first Executive Officer. Henry, a renowned experimental physicist, continued the practice of a federal agency directly supporting research through grants to individual investigators to pursue fundamental research. Also, the Institution represented a base for external support of scientific and engineering research since, during the 1850s, about 100 academic institutions were established with science and engineering emphases.

Thus, the pendulum had made one complete swing in the hundred years since the signing of the Constitution. In the early years, the government viewed itself as having no more than an indirect influence on the development of science and technology, but over time its role changed from indirect to direct. This change was justified in large part because advances in science and technology came to be viewed as critical in promoting the public interest.

TOWARD A NATIONAL SCIENCE AND TECHNOLOGY INFRASTRUCTURE

Scientists had long looked toward the European universities for training in the sciences, but now an academic infrastructure was beginning to develop in the United States. Harvard University awarded its first bachelors of science degree in 1850.

The development of an academic science base and the birth of technology-based industries (e.g., the electrical industry) in the late 1850s established what would become the foundation for America's technological preeminence.

In 1863, during the Civil War, Congress established the National Academy of Sciences. The federal government funded the Academy but not the members affiliated with it who had (Unesco 1968, p. 14):

> an obligation to investigate, examine, experiment, and report upon any subject of science or art in response to a request from any department of the Government.

Then, as today, the Academy is independent of governmental control.

The Morrill Act of 1862 established the land grant college system thereby formally recognizing the importance of trained individuals in the agricultural sciences. The Act charged each state to establish at least one college in the agricultural and mechanical sciences. Each state was given 30,000 acres of federal land per each elected U.S. Senator and Representative. An important outgrowth of this land grant system was a mechanism or infrastructure through which state and federal governments could financially support academic research interests.

Although the government was encouraging an infrastructure to support science and technical research, it did not have a so-called in-house staff of permanent professionals who were competent to identify either areas of national importance or areas of importance to specific agencies. In 1884, Congress established the Allison Commission to consider this specific issue. While many solutions were debated, including the establishment of a Department of Science—an idea that resurfaces every few decades—the Commission soon disbanded without making any recommendations much less reaching closure on the matter. One could conclude from the inaction of the Commission that it favored the decentralized administrative architecture that had evolved over time as opposed to a centralized one.

TOWARD AN INDUSTRIAL SCIENCE AND TECHNOLOGY INFRASTRUCTURE

Most scientists in the United States in the 1870s and 1880s had been trained in Europe, Germany in particular. What they experienced firsthand were the strong ties between European industries and graduate institutions of learning. Companies invested in professors and in their graduate students by providing them with funding and access to expensive materials and instruments, and in return the companies gained lead-time toward new discoveries as well as early access to the brightest graduate students as soon as they completed their studies. This form of symbiotic arrangement became the norm for the European-trained scientists who were working in U.S. industries and U.S. universities toward the end of the century.

By the turn of the century, it was widely accepted among industrial leaders that scientific knowledge was the basis for engineering development and it was the key to remaining competitive. Accordingly, industrial research laboratories soon began to blossom as companies realized their need to foster scientific knowledge outside of the university setting. There are a number of examples of this strategy. (Hounshell (1996) provides an excellent history of the growth of U.S. industrial research organizations.)

General Electric (GE) established the General Electric Research Laboratory in 1900 in response to competitive fears that improved gas lighting would adversely affect the electric light business, and that other electric companies would threaten GE's market share as soon as the Edison patents expired.

Similarly, AT&T was at the same time facing increasing competition from radio technology. In response, AT&T established Bell Laboratories to research new technology in the event that wire communications were ever challenged.

And as a final example, Kodak realized at the turn of the century that it must diversify from synthetic dyes. For a number of years Kodak relied on German chemical technology, but when that technology began to spill over into other areas, such as photographic chemicals and film, Kodak realized that their competitive long-term health rested on their staying ahead of their rivals. Kodak too formed an in-house research laboratory.

Many smaller firms also realized the competitive threats that they could potentially face as a result of technological competition, but because of their size they could not afford an in-house facility. So as a market response, contract research laboratories began to form. Arthur D. Little was one such contract research laboratory that specialized in the area of chemicals.

Just as industrial laboratories were growing and being perceived by those in both the public and private sectors as vitally important to the economic health of the nation, private foundations also began to grow and to support university researchers. For example, the Carnegie Institution of Washington was established in 1902, the Russell Sage Foundation in 1907, and the Rockefeller Foundation in 1913.

In the early-1900s science and technology began to be embraced—both in concept and in practice—by the private sector as the foundation for long-term competitive survival and general economic growth.

WORLD WAR I AND THE YEARS THAT FOLLOWED

Increased pressure on the pace of scientific and technical advancements came at the beginning of World War I. The United States had been cut off from its European research base. Congress, in response, established the Council of National Defense in 1916 to identify domestic pockets of scientific and technical excellence.

The National Academy of Sciences recommended to President Woodrow Wilson the formation of the National Research Council to coordinate cooperation between the government, industry, and the academic communities toward common national goals. The Allison Commission had failed in 1884 to formulate an infrastructure to undertake this task.

The prosperity of the post-World War I decade also created an atmosphere supportive of the continued support of science and technology.

In 1920, there were about 300 industrial research laboratories, and by 1930 there were more than 1,600. Of the estimated 46,000 practicing scientists in 1930, about half were at universities and over a third were in industry. Herbert Hoover was Secretary of Commerce at this time. He adopted the philosophy that (Unesco 1968, p. 18):

> pure and applied scientific research constitute a foundation and instrument for the creation of growth and efficiency of the economy.

Two important events occurred in 1933 in response to the Great Depression and the subsequent national economic crisis. One event was the appointment of a Science Advisory Board, and the other event was the establishment of a National Planning Board. Whereas the National Research Council had been organized around *fields of science* to address governmental needs, the Science Advisory Board was multi-field and organized around *impending national problems*. The National Planning Board was formed on the presumption that there were areas of economic concern that required a national perspective rather than a field-of-science perspective.

In 1934, the National Resources Committee replaced the National Planning Board, and the Committee then subsumed the Science Advisory Board. The bottom line was, after all of the organizational issues were settled, that the federal government recognized through the formation of these committees and boards that it had and would continue to have an important coordinating role to play in the science and technology planning toward a national goal of economic well being. Hence, the pendulum began to swing again, this time away from government having a hands-on role toward it having an indirect influence on planning the environment for science and technology.

In 1938, the Science Committee of the National Resources Committee issued a multi-volume report entitled, *Research—A National Resource*. Some important first principles were articulated in that report. These principles have since then formed a basis for economists and policy makers to rationalize and justify the role of government in science and technology. The report is explicit that:

- There are certain fields of science and technology which the government has a Constitutional responsibility to support. These fields include defense, determination of standards, and certain regulatory functions.

- The government is better equipped to carry on research in certain fields of science than the private sector. These are areas where "research is unusually costly in proportion to its monetary return but is of high practical or social value" (p. 25). Examples cited in the report include aeronautical and geological research.
- Research by the government "serves to stimulate and to catalyze scientific activity by nongovernmental agencies. In many fields, new lines of research are expensive and returns may be small or long delayed. Industry cannot afford to enter such fields unless there is reasonable prospect of definite financial gain within a predictable future, and it is under such circumstances that the Government may lead the way." (p. 26). One example cited was the Navy Department's influence on the development of the steel industry.

WORLD WAR II AND THE YEARS THAT FOLLOWED

The involvement of the United States in World War II had a dramatic impact on the scope and direction of government's support of science and technology. Prior to the war, there were about 92,000 scientists, with about 20 percent in government and the remaining 80 percent being almost equally divided between universities and the more than 2,200 industrial laboratories. Clearly, the United States had a significant scientific resource base to draw upon for its war efforts.

In 1940, President Franklin D. Roosevelt established the National Defense Research Committee, and he asked Vannevar Bush, President of Carnegie Institution of Washington, to be its chairman. The purpose of this committee was to organize scientific and technological resources toward enhancing national defense. It soon became apparent that this task required an alternative administrative structure.

In 1941, Roosevelt issued an Executive Order establishing the Office of Scientific Research and Development (OSRD) with Bush as Director. The OSRD did not conduct research, rather it realized that there were pockets of scientific and technological excellence throughout the country, and through contractual relationships with universities and industry and government agencies it could harness national strengths with a focus on ending the war. One hallmark event from the efforts of the OSRD was the establishment of the Los Alamos Laboratory in New Mexico under the management of the University of California. What came about from the collective efforts of the resources acquired by the Office were not only atomic weapons but also radar.

It was clear by 1944 that World War II was almost over. President Roosevelt then asked Bush to develop recommendations as to how scientific advancements could contribute in the larger sense to the advancement of national welfare. In his November 17, 1944 letter to Bush, President Roosevelt stated:

> The Office of Scientific Research and Development, of which you are the Director, represents a unique experiment of team-work and cooperation in coordinating scientific research and in applying existing scientific knowledge to the solution of the technical problems paramount in war. ... There is ... no reason why the lessons to be found in this experiment cannot be profitably employed in times of peace. This information, the techniques, and the research experience developed by the Office of Scientific Research and Development and by the thousands of scientists in the universities and in private industry, should be used in the days of peace ahead for the improvement of the national health, the creation of new enterprises bringing new jobs, and the betterment of the national standard of living. ... New frontiers of the mind are before us, and if they are pioneered with the same vision, boldness, and drive with which we have waged this war we can create a fuller and more fruitful employment and a fuller and more fruitful life.

Shortly before asking Bush to prepare this report, Senator Harley M. Kilgore from West Virginia had introduced a bill to create a National Science Foundation. The Kilgore bill recommended giving authority to federal laboratories to allocate public moneys in support of science to other government agencies and to universities. Clearly, this recommendation gave a direct role to government in shaping the technological course of the country not only in terms of scientific direction but also in terms of what groups would conduct the underlying research. The bill was postponed until after the war.

Bush submitted his report, *Science—the Endless Frontier*, to President Roosevelt on July 25, 1945. In Bush's transmittal letter to the president he stated:

> The pioneer spirit is still vigorous within this Nation. Science offers a largely unexplored hinterland for the pioneer who has the tools for his task. The reward of such

> exploration both for the Nation and the individual are great. Scientific progress is one essential key to our security as a nation, to our better health, to more jobs, to a higher standard of living, and to our cultural progress.

The foundations set forth in *Science—the Endless Frontier* are:

- "Progress ... depends upon a flow of new scientific knowledge" (p. 5).
- "Basic research leads to new knowledge.[1] It provides scientific capital. ... New products and new processes do not appear full-grown. They are founded on new principles and new conceptions, which in turn are painstakingly developed by research in the purest realms of science" (p. 11).
- "The responsibility for the creation of new scientific knowledge ... rests on that small body of men and women who understand the fundamental laws of nature and are skilled in the techniques of scientific research" (p. 7).
- "A nation which depends upon others for its new basic scientific knowledge will be slow in its industrial progress and weak in its competitive position in world trade, regardless of its mechanical skill" (p. 15).
- "The Government should accept new responsibilities for promoting the flow of new scientific knowledge and the development of scientific talent in our youth" (p. 7).
- "If the colleges, universities, and research institutes are to meet the rapidly increasing demands of industry and Government for new scientific knowledge, their basic research should be strengthened by use of public funds" (p. 16).
- "Therefore I recommend that a new agency for these purposes be established" (p. 8).

Bush recommended in his report the creation of a National Research Foundation. Its proposed purposes were to:

> develop and promote a national policy for scientific research and scientific education, ... support basic research in nonprofit organizations, ... develop scientific talent in American youth by means of scholarships and

[1] The term "basic research" is credited to Vannevar Bush. He proffered the definition: "Basic research is performed without thought of practical ends."

> fellowships, and ...contract and otherwise support long-range research on military matters.

Bush envisioned a National Research Foundation that would provide funds to institutions outside government for the conduct of research. Thus, this organization differed from Kilgore's proposed National Science Foundation in that Bush advocated an indirect role for government. There was agreement throughout government that an institutional framework for science was needed, but the nature and emphases of that framework would be debated for yet another five years.

Science—the Endless Frontier affected the scientific and technological enterprise of this nation in at least two ways. It laid the basis for what was to become the National Science Foundation in 1950. Also, it set forth a paradigm that would over time influence the way that policy makers and academic researchers thought about the process of creating new technology. The so-called linear model set forth by Bush is often represented by:

Basic Research → Applied Research → Development → Enhanced Production → Economic Growth

Complementing *Science—the Endless Frontier* was a second, and often overlooked, report prepared in 1947 by John Steelman, then Chairman of the President's Scientific Research Board. As directed by an Executive Order from President Harry Truman, Steelman, in *Science and Public Policy*, made recommendations on what the federal government could do to meet the challenge of science and assure the maximum benefits to the Nation. Steelman recommended that national R&D expenditures should increase as rapidly as possible, citing (p. 13):

> 1. Need for Basic Research.
> Much of the world is in chaos. We can no longer rely as we once did upon the basic discoveries of Europe. At the same time, our stockpile of unexploited fundamental knowledge is virtually exhausted in crucial areas.
> 2. Prosperity.
> This Nation is committed to a policy of maintaining full employment and full production. Most of our frontiers have disappeared and our economy can expand only with more intensive development of our present resources. Such expansion is unattainable

without a stimulated and growing research and development program.

3. International Progress.
The economic health of the world—and the political health of the world—are both intimately associated with our own economic health. By strengthening our economy through research and development we increase the chances for international economic well-being.
4. Increasing Cost of Discovery.
The frontiers of scientific knowledge have been swept so far back that the mere continuation of pre-war growth, even in stable dollars, could not possibly permit adequate exploration. This requires more time, more men, more equipment than ever before in industry.
5. National Security.
The unsettled international situation requires that our military research and development expenditures be maintained at a high level for the immediate future. Such expenditures may be expected to decrease in time, but they will have to remain large for several years, at least.

An important element of the Steelman report was the recommended creation of a National Science Foundation, similar in focus to the National Research Foundation outlined by Bush. And, Congress passed the National Science Foundation Act in 1950.

Renewed post-war attention toward science and technology came with the success of the Soviet Union's space program and the orbit of its Sputnik I in October 1957. In response, President Dwight D. Eisenhower championed a number of committees and agencies to ensure that the United States could soon be at the forefront of this new frontier. Noteworthy was the National Defense Education Act of 1958, which authorized $1 billion in federal moneys for support of science, mathematics, and technology graduate education. This proposal is precisely the type of support that Bush recommended in his report.

As the post-World War II period came to close, there was a well-established national and industrial infrastructure to support the advancement of science and technology. But, more important than the infrastructure, there was an imbedded belief that scientific and technological advancements are fundamental for economic growth, and that the government has an important supporting role—both direct and indirect—to ensure such growth.

Every president since President Eisenhower has initiated major science and technology policy initiatives:

- President John F. Kennedy set the goal of sending a man to the moon by the end of the 1960s and funded the needed programs to make this a reality.
- President Lyndon B. Johnson emphasized the use of scientific knowledge to solve social problems through, for example, his War on Poverty.
- President Richard M. Nixon dramatically increased federal funding for biomedical research as part of his War on Cancer.
- President Gerald R. Ford created the Office of Science and Technology Policy (OSTP) within the Executive Branch.
- President James E. Carter initiated research programs for renewable energy sources such as solar energy and fission.
- During President Ronald W. Reagan's administration, expenditures on defense R&D increased dramatically as part of his Star Wars system.
- President George Bush (not related to Vannevar Bush) set forth this nation's first technology policy (see below) and increased the scope of the National Institute of Standards and Technology (NIST).
- President William J. Clinton established important links between science and technology policy, championing programs to transfer publicly-funded technology to the private sector.
- And, President George W. Bush advocated the permanency of the R&E tax credit, as discussed in the following chapters.

3 ECONOMIC GROWTH AND TECHNOLOGICAL CHANGE

*E*conomic growth is typically measured at the macroeconomic level in terms of Gross Domestic Product (GDP). Figure 3.1 presents GDP per capita estimates for several countries. Note that GDP per capita is highest in the United States compared to any of the other major industrial nations shown in the figure (France, Germany, Japan, and the U.K.).

This pattern raises two questions. One, Why do economies grow? And two, Why has the U.S. economy out performed that of other industrial nations even after controlling for national size?

Economic growth is typically examined based on what economists refer to as a production function framework. Simply put, a production function represents the relationship between the output of an economic unit (a plant, firm, industry, or economy) and the factors of production—or inputs or resources—used to produce that output.

If output, Q, can be defined in terms of the two most basic factors of production, the stock of capital (plant and equipment), K, and the stock of labor, L, then a basic production function can be written as:

$$(1) \qquad Q = f(K,L).$$

Equation (1) simply represents that a firm's, industry's, or economy's output will change in response to changes in either the quantity or quality of capital or the quantity or quality of labor. However, to account for other influences on output such as new technology, the production relationship in equation (1) can be modified most simply to include a catch-all variable, A, as:

$$(2) \qquad Q = A\,F(K,L)$$

where, to be more specific, A is a shift factor to account for exogenous technological factors (as opposed to conventional factor inputs such as K and L) that affect production, and where, because of the inclusion of A in equation (2), F(K,L) is distinct from f(K,L) in equation (1).

By dividing both sides of equation (2) by the combination of K and L inputs (i.e., by total factors) denoted by F(K,L), variable A can be interpreted as an index of output per unit of all inputs or of total factor productivity (TFP). See Figure 3.2 for the trend in TFP in the U.S. non-farm sector.

(3) $A = Q / F(K,L)$

Early on, Solow (1957), who was subsequently awarded the Nobel Prize in economics, estimated a variation of equation (3) using aggregate U.S. data. He calculated percentage changes in the value of A between 1909 and 1949. His analysis showed that more than 87 percent of the growth in the U.S. economy could not be explained by the growth in capital and labor, and hence, the residual or unexpected portion of growth must be attributable to something else. Solow speculated that what was captured in his residual calculation may reflect technology advance over time. Changes in A from equation (3) measure what is called total factor productivity growth, or technological advancement.

Other researchers, using alternative frameworks of analysis, independently reached similar conclusions as Solow. Abramovitz (1956), for example, referred to the unexplained portion of growth more cautiously as a measure of our ignorance, meaning that at that time economists were able to calculate unexplained growth but were not able to offer a full explanation for what caused it. Unlike Solow, Abramovitz speculated in some detail that growth not attributable to capital and labor was likely due to improvements in education and increases in R&D.

The academic literature is replete with theories to explain growth over time. The so-called old growth theory literature is based on more sophisticated versions of equation (1). That is, this literature emphasizes additional inputs aside from K and L such as investments in R&D and education. As well, it emphasizes the greater specificity by which inputs are measured including consideration for the heterogeneity of K and L (e.g., new vintages of K embody others' technological investments).

Figure 3.1. Real GDP per Capita for Selected Countries, 1960-1996 (U.S. $1000)

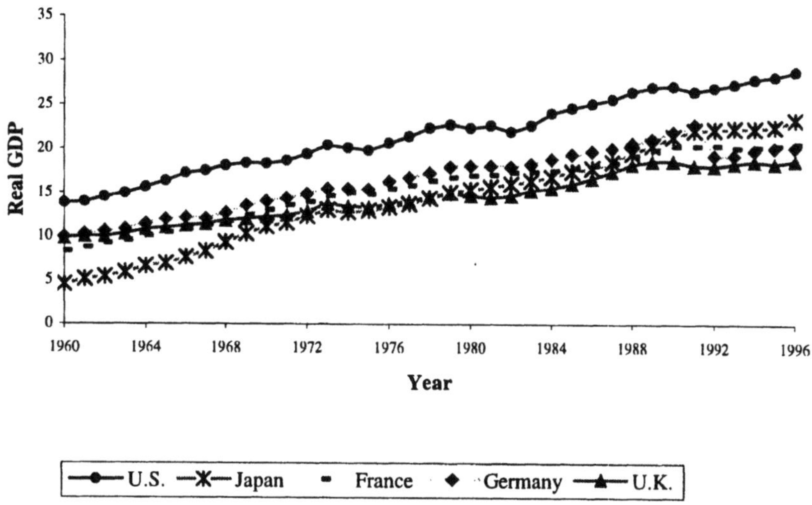

Source: *Science and Engineering Indicators–2000*, Appendix Table 7-3.

"New" growth theory emphasizes the influence of other factors on growth that are not directly specified in an expanded version of equation (1). These factors include, for example, technologies or efficiencies that spillover into a firm's production function either from other firms or from general advances in the economy (such as information technology), or that spillover into a nation's production function from trade policies. New growth theory is also based on highly stylized analytical models, which account for the incentives of economic agents to invest in new technology.

Figure 3.3 expands upon Figure 1.1 to incorporate these ideas. In particular, two new elements are included in Figure 3.3 that were not in Figure 1.1. First, in-house or private investment in R&D is shown to have an influence on how the firm responds—its entrepreneurial response—to the interaction of its strategic direction and its competitive environment. Second, we have capsulated the essence of new growth theory by simply acknowledging that external influences affect firm performance directly as well as indirectly through innovation and the incentive structures associated with the innovation process.

Figure 3.2. U.S. Private Non-Farm TFP, 1948-1997

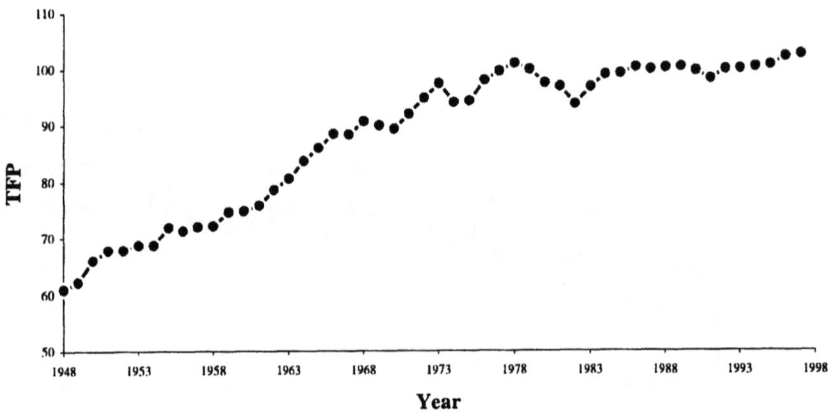

Source: U.S. Bureau of Labor Statistics

Regardless of whether one adheres to the more narrow old theories or the broader new theories, the evidence is overwhelming that technology drives economic growth. There was renewed interest in promoting economic growth in the postwar aftermath of the destruction of the industrial base of many nations. Thus, it is not surprising that greater attention was devoted to the analysis of R&D, which was hypothesized early on to be an important determinant of economic growth. And as well, current innovation policy initiatives, both in the United States and in most developed nations, focus heavily on incentives to stimulate private-sector R&D.

The following chapter considers important dimensions of R&D.

Figure 3.3. The Entrepreneurial Process: A Second Look

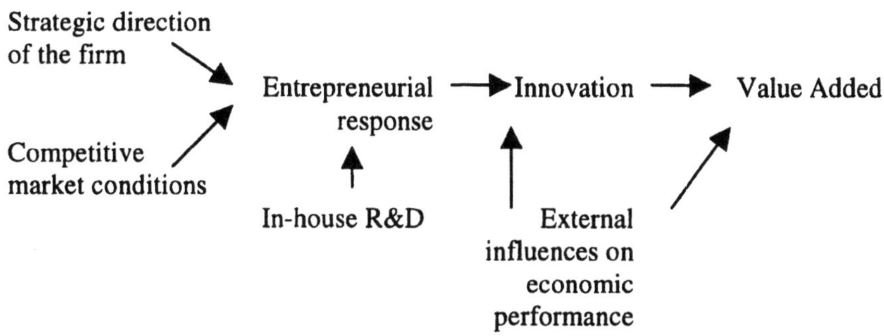

4 DIMENSIONS OF R&D

For purposes of measurement, there are three fundamental dimensions of R&D. The first relates to the *source* of funding of R&D (who finances the investment), the second to the *performance* of R&D (who conducts the research and development), and the third to the *character of use* of R&D (whether the undertaking is of a basic or applied nature, or development). These three fundamental dimensions are not mutually exclusive.

There are also three important aspects of R&D. One aspect is the relationship between the size of firms and its propensity to conduct R&D, the second is the geographical distribution of the R&D, and the third is the relationship between R&D and total factor productivity growth.

SOURCES OF FUNDING OF R&D

The top row of Table 4.1 shows the sources for the $247.0 billion of R&D expenditures in the United States in 1999. Industry accounted for nearly 69 percent of those expenditures; the federal government another 27 percent; and all other sources, including state and local governments, universities and colleges, and other nonprofit institutions, about five percent.

The primacy of industry in funding R&D has not always held, as shown in Figure 4.1. In the aftermath of World War II, up through the early 1980s, the federal government was the leading provider of R&D funds in the nation. Although a federal R&D presence was in existence before then, during the war the federal government dramatically expanded its R&D effort by establishing a network of federal laboratories, including atomic weapons laboratories. It was at that time that the federal government also greatly increased its support to extramural R&D performers, especially to a select group of universities and large industrial firms. After the war (along with the widespread influence of the Bush and Steelman reports), federal R&D support continued to expand for both defense and non-defense purposes, including health R&D in the National Institutes of Health and—after the establishment of the National Science Foundation in 1950—a broad portfolio of fundamental research activities. As a result of a post-Sputnik national commitment to catch up to the Soviet space successes, federal support for space-related R&D mushroomed in the late 1950s and early 1960s. By 1960, the federal government accounted for 65 percent of the nation's total investment (80 percent of which was for defense), and industry accounted for 32 percent of the total.

Table 4.1. National R&D Expenditures, by Performer and Funding Source, 1999 ($Millions)

	Source of R&D Funds						
Performer	Federal govt.	Industry	Universities and colleges	Non-federal govt.	Other nonprofit institutions	Total	Percent distribution, by performer
TOTAL R&D	65,853	169,312	5,838	2,085	3,912	247,000	100.0%
Federal govt.	17,362	..	..	..	..	17,362	7.0%
Industry	19,937	165,955	..	..	..	185,892	75.3%
Industry FFRDCs	2,166	..	..	..	..	2,166	0.9%
Universities and colleges	16,137	2,163	5,838	2,085	2,032	28,255	11.4%
University FFRDCs	6,169	..	..	..	..	6,169	2.5%
Other nonprofit institutions	3,246	1,194	..	..	1,880	6,320	2.6%
Nonprofit FFRDCs	836	..	..	..	..	836	0.3%
Percent distribution, by source	26.7%	68.5%	2.4%	0.8%	1.6%	100%	

Source: National Science Board (2000).

Figure 4.1. U.S. R&D Funding by Source, 1953-1998
(Billions $1992)

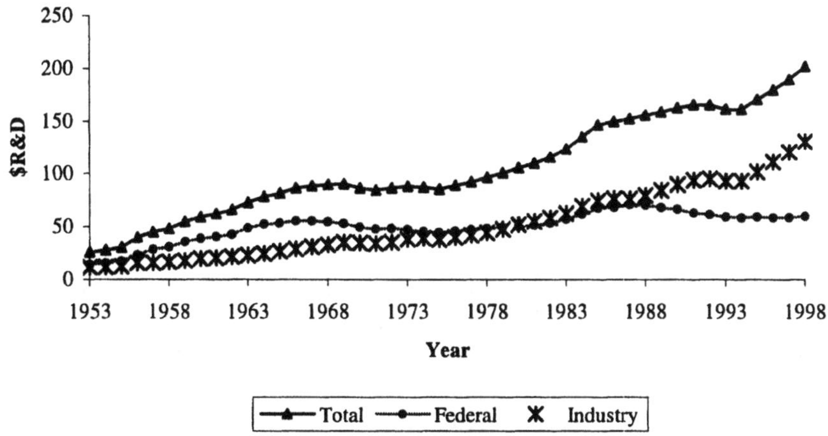

Source: National Science Board (2000, Appendix Table 2-6).

Over the next twenty years the federal government continued to be the leading source of R&D funding, although the direction of its funding shifted over time. In the early 1960s, the relative defense share of federal R&D funding dropped precipitously from 80 percent in 1960 to about 50 percent in 1965, where it fluctuated narrowly until 1980. Early on, R&D for space exploration was the primary non-defense recipient of federal R&D funding. Indeed more than three-fourths of federal non-defense R&D funds were in support of NASA's mission activities by 1965.

By 1970, however, after the success of several lunar landings, support for other non-defense purposes began to claim an increasingly larger share of the federal R&D totals, and continued to do so throughout the 1970s; notably growth in federal energy R&D occurred as a response to the several oil embargoes. Also by 1970, R&D support from industry was on the rise, and it accounted for just over 40 percent of the total national R&D effort. As a result of relatively flat federal funding in the 1970s and continual slow growth from the industrial sector, the federal government and industry accounted for about equal shares by the early 1980s.

Since then, the federal government's share of R&D decreased to about 40 percent of total in 1990 to its current share, somewhat below 30 percent. Initially, the decreasing federal share came about even though federal dollar support for R&D—in absolute terms—was increasing.

Between 1980 and 1987, federal R&D rose about 40 percent after adjusting for inflation. Most of this growth, however, was in support of defense activities so that by 1987, the defense R&D share had grown to two-thirds of the federal R&D total (its highest share since 1963). After the break-up of the Soviet Union, the imperative for continual growth in federal defense R&D support was not as strong and the federal R&D total once again slowed (and even fell in constant dollars).

In terms of which agencies provide the R&D funds, federal sources are highly concentrated among just a few agencies. According to the latest data provided by the agencies themselves, of the $74 billion obligated for R&D *and* R&D plant in fiscal year 1998, just five accounted for 94 percent of all funds: Department of Defense (48%), Department of Health and Human Services, primarily the National Institutes of Health (19%), National Aeronautics and Space Administration (13%), Department of Energy (9%), and National Science Foundation (3%).

Concurrent with recent reductions in federal R&D spending, major changes have also transpired in industrial R&D spending patterns. After lackluster funding in the early 1990s (reflecting the impact of mild recessions on its R&D activities) industry R&D support has grown rapidly since 1994 and now accounts for almost 70 percent of the national R&D total. As a result, and compared with the funding patterns of the mid-1960s, industry and government have reversed positions.

R&D PERFORMERS

R&D is performed in what has been called the U.S. national innovation system (Nelson 1993). The system's many dimensions have been studied by many; according to Crow and Bozeman (1998, p. 42), the U.S. national innovation system may be thought of as:

> the complex network of agents, policies, and institutions supporting the process of technical advance in an economy.

The venue for the performance of R&D within the system—and this is true in all industrial nations—are research laboratories. The laboratory performers of R&D correspond to the sectors that finance R&D, but not all R&D funded by a sector is performed in that sector.

For example, industry performed in its laboratories approximately $186 billion of R&D in 1999, of which $166 billion came from industry itself. The additional amount of R&D performed by industry came from the federal government.

Almost one-third of the R&D funded by the federal government is performed in industry, and more than one-half of these funds are spent in the aircraft, missiles, and transportation equipment industries. Universities and colleges fund only about 20 percent of the R&D they perform in their laboratories. Fifty-seven percent of R&D performed in universities and colleges comes from the federal government and the rest equally from industry, nonprofit institutions, and nonfederal government sources.

Figure 4.1 shows that since the late-1980s the federal government has decreased its funding of national R&D. The lion's share of that decrease has come in the form of federal allocations for R&D performed in industry, for which the R&D level of support displays a somewhat roller-coaster-like pattern. The latest peak in federal support for industrial R&D was due to major defense-related funding increases for President Reagan's Strategic Defense Initiative, prior to the collapse of the Soviet Union. By contrast, federal funding to universities and colleges, adjusted for inflation, has increased slightly each year since at least the late 1970s.

There are other important dimensions to the performance of industrial R&D. Approximately three-fourths of total industrial R&D is performed in manufacturing industries. The dominant manufacturing industries in terms of dollars of R&D performed are chemicals and allied products, electrical equipment (including computers), and transportation equipment. The remaining one-fourth is performed in the nonmanufacturing sector, including services. Computer-related services are the leaders therein. The steep growth of R&D performed in the services is a relatively recent phenomenon. As recently as 15 years ago, manufacturers still accounted for more than 90 percent of total industrial R&D.

Also, not all industry-performed R&D occurs within the geographical boundaries of the United States. Of the nearly $186 billion in R&D performed by industry in 1998 (the latest year for which the foreign-performed data are available), about $16 million, or about nine percent, were conducted in other countries.

Foreign investments in R&D are not unique to U.S. firms; the outflow of U.S. industrial R&D into other countries is approximately offset by an inflow of others' R&D to be performed in the United States. Most (68%) of U.S.-funded R&D abroad was performed in Europe—primarily in Germany, the United Kingdom, and France. The current European share of U.S. industry's offshore R&D activity, however, is somewhat less than the 75 percent share reported for 1982 (peak year). Overall, U.S. R&D

investments abroad have generally shifted away from the larger European countries and Canada, and toward Japan, several of the smaller European countries (notably Sweden and the Netherlands), Australia, and Brazil. Pharmaceutical companies accounted for the largest industry share (18 percent of U.S. 1997 overseas R&D), which was equivalent to 21 percent of their domestically-financed R&D. Much of this pharmaceutical R&D took place in the United Kingdom.

Substantial R&D investments are made by foreign firms in the United States. From 1987 to 1996, inflation-adjusted R&D growth from majority-owned affiliates of foreign firms averaged 10.9 percent per year, and are now roughly equivalent to U.S. companies' R&D investment abroad. Affiliates of firms domiciled in Germany, Switzerland, the United Kingdom, France, and Japan collectively account for 72 percent of this foreign funding.

Foreign-funded R&D in the United States in 1996 was concentrated in drugs and medicines (mostly from Swiss, German, and British firms), industrial chemicals (funded predominantly by German and Dutch firms), and electrical equipment (one-third of which came from French affiliates).

R&D BY CHARACTER OF USE

Vannevar Bush is credited for first using the term "basic research." In his 1945 report to President Roosevelt, *Science—the Endless Frontier*, Bush used the term and defined it to mean research conducted without thought of practical ends. Since then, policy makers have been concerned about definitions that appropriately characterize the various aspects of scientific inquiry that broadly fall under the label of R&D and that relate to the linear model that Bush proffered.

Definitions are important to the National Science Foundation because it collects expenditure data on R&D. For those data to accurately reflect industrial and academic investments in technological advancement, and for those data to be comparable over time, there must be a consistent set of reporting definitions.

The classification scheme used by the National Science Foundation for reporting purposes was developed for its first industrial survey in 1953-1954, as documented in Link's (1996b) history of the classification scheme. While minor definitional changes were made in the early years, namely to modify the category originally referred to as "basic or fundamental research" to simply "basic research," the concepts of basic research, applied research, and development have remained much as was implicitly contained in Bush's 1945 linear model:

Basic Research → Applied Research → Development

The objective of *basic research* is to gain more comprehensive knowledge or understanding of the subject under study, without specific applications in mind. Basic research is defined as research that advances scientific knowledge but does not have specific immediate commercial objectives, although it may be in fields of present or potential commercial interest. Much of the scientific research that takes place at universities is basic research.

Applied research is aimed at gaining the knowledge or understanding to meet a specific recognized need. Applied research includes investigations oriented to discovering new scientific knowledge that has specific commercial objectives with respect to products, processes, or services.

Development is the systematic use of the knowledge or understanding gained from research directed toward the production of useful materials, devices, systems, or methods, including the design and development of prototypes and processes.

Approximately 61 percent of national R&D is development, with 23 percent of R&D being allocated to applied research and 16 percent being allocated to basic research. Different sectors contribute disproportionately to the nation's funding and performance of these R&D component categories. Applied research and development activities are primarily funded by industry and performed by industry. Basic research, however, is primarily funded by the federal government and generally performed in universities and colleges.

The decline in federal support of R&D over the past decade has primarily come at the expense of applied research and development performed in industry. The decline in federal support of R&D over the past decade has primarily come at the expense of applied research and development performed in industry.

R&D ACTIVITY IN LARGE AND SMALL FIRMS [1]

Table 4.2 reports the level of R&D expenditures, from all sources, for the largest R&D-performing firms in the United States in 1997. Several patterns can be seen from the data in this table:

---

[1] A portion of this section comes directly from the contribution of David Audretsch to Audretsch et al. (2002).

- Microsoft (the tenth-largest R&D-active company) invests about one-fourth of the amount of R&D invested by General Motors (the largest R&D-active company). Thus, even among the R&D giants, R&D expenditures vary dramatically.
- Microsoft is, however, more than three times as R&D intensive as General Motors, meaning that it invests nearly three times the amount as General Motors relative to its sales.
- In general (with notable exceptions such as Cabletron Systems), larger R&D performers also spend more on R&D relative to their size than do the lower-ranked firms in the list of the top 100.
- The level of R&D expenditures is not unrelated to the industry of the R&D performer given that technological opportunities vary across sectors. General Motors and Ford Motor Company are in the transportation industry. IBM, Lucent Technologies, Hewlett-Packard, Motorola, and Intel are in information systems, and Microsoft supports aspects of that industry.

Firm-specific data on R&D expenditures for small-sized companies are not readily available. However, aggregate NSF data show that $98 billion of the $169 billion that industry spent on R&D in 1998 was performed in firms with 10,000 or more employees. Thirty billion dollars (18 percent of the industry total) was performed in firms with fewer than 500 employees. Similarly, 70 percent of the funds expended were part of company R&D budgets that exceeded $100 million.

Stylized facts aside, there is a more subtle and perhaps more important R&D-related issue. Since the early 1980s, policy makers have been concerned that American firms in key industries were losing their dominance of global markets. During the 1990s, these same industries seem to have reemerged as major international competitors. While some of this resurgence is a response to purposive policies, a portion of it can also be attributable to small firms, many of which were not in existence in the early 1980s to be affected by policy and many of which do not even conduct R&D. Still, during the 1990s, small firms were a driving engine of growth, job creation, and renewed global competitiveness through innovation.

Table 4.2. Largest R&D-Active U.S. Companies

Rank in 1997	Company	$R&D (millions)	$R&D/$Sales (%)
1	General Motors	8,200.0	4.9
2	Ford Motor Company	6,327.0	4.1
3	IBM	4,307.0	5.2
4	Lucent Technologies	3,100.6	11.8
4	Hewlett-Packard	3,078.0	7.2
5	Motorola	2,748.0	8.6
7	Intel	2,347.0	9.4
8	Johnson & Johnson	2,140.0	9.5
9	Pfizer	1,928.0	15.4
10	Microsoft	1,925.0	16.9
.			
.			
.			
95	Imation	194.9	8.9
96	Dana	193.0	2.2
97	Thermo Electron	191.6	5.4
98	Eastman Chemical	191.0	4.1
99	Cabletron Systems	181.8	13.2
100	Whirlpool	181.0	2.1

Source: National Science Board (2000, Appendix Table 2-58).

There is a rich literature related to the performance of R&D in small firms as compared to large firms. Some of the conclusions from this research are:

- Large firms have a greater propensity to patent than small firms.
- Small firms are just as innovative as large firms, in general. However, in some industries, large firms have the innovative advantage (pharmaceuticals, aircraft), while in other industries small firms have the innovative advantage (software, biotechnology).
- Small-firm and large-firm innovative activities are complementary.

Table 4.3 provides a selected summary of the findings from the literature on innovation and firm size.

The economic importance of small firms, including the innovative differences between small firms and large firms, requires an explanation since the share of overall economic activity attributable to small firms is small and it did not increase during the 1990s. The explanation relevant to the focus of this book begins with a model of the knowledge production function first set forth by Griliches (1979).

The simplified production function in equation (1) above can be expanded conceptually and analytically to include the stock of knowledge as a discrete input along with K and L. One investment in knowledge that many firms make is in R&D. However, there are other key factors that generate knowledge for the firm besides R&D, and in fact many small firms do not even conduct R&D yet they are very innovative. Some such firms rely on knowledge that spills over from external sources including universities, and small firms are relatively more adept at absorbing knowledge from external sources than large firms. Table 4.4 provides a brief summary of this spillover literature with respect to small firms.

Table 4.3. Selected Studies of the Relationship Between Innovation and Firm Size

Innovation Measure	Findings	Authors
R&D	R&D spending is positively related to firm size	Mueller (1967) Grabowski (1968) Mansfield (1968)
Patents	Patenting is positively or proportionally related to firm size	Scherer (1965, 1983) Pakes and Griliches (1980) Hall, Griliches, and Hausman (1986) Schwalback and Zimmermann (1991)
New product innovations	Parity across firm size, although there are differences according to industry	Acs and Audretsch (1990) Audretsch (1995)
Adoption of advanced manufacturing technologies	Positive relationship between firm size and the probability of adopting an advanced manufacturing technology	Romeo (1975) Dunne (1994) Siegel (1999)

Included in Table 4.4 under the source category of individual spillovers—and economists and policy makers use the term spillover to refer to the diffusion of an economic good or economic activity, either intentionally or unintentionally due to market failure—are new employees. An interesting question is: Why are small firms able to exploit knowledge embodied in new employees to a greater extent than large firms? One answer is that new and small firms provide the opportunity for creative individuals to implement new ideas that otherwise would be rejected or would remain unexploited in an organizationally rigid firm. New firms thus serve as agents of change. In a global economy where comparative advantage is based in large part on innovation, small firms are a critical resource.

Public policies to enhance innovation in small firms are discussed in Chapter 10.

R&D ACTIVITY BY GEOGRAPHIC LOCATION

R&D activity in the United States is highly concentrated in a small number of states. In 1997, the 20 highest-ranking states in R&D accounted for about 86 percent of the U.S. total; the lowest 20 states accounted for only four percent. California, at nearly $42 billion, had the highest level of R&D expenditures; it alone accounted for approximately one-fifth of the $199 billion U.S. total. The six states with the highest levels of R&D expenditures—California, Michigan, New York, New Jersey, Massachusetts, and Texas (in decreasing order of magnitude)—accounted for nearly two-thirds of the national effort. Among these top ten states, California's R&D effort exceeded, by nearly a factor of three, the next-highest state, Michigan, with $14 billion in R&D expenditures. After Michigan, R&D levels declined relatively smoothly to approximately $7 billion for Maryland.

States that are national leaders in total R&D performance are usually ranked among the leading sites in industrial and academic R&D performance. For industrial R&D, nine of the top ten states were among the top ten for total R&D, with Ohio of the top industrial R&D states replacing Maryland. For academic R&D, North Carolina and Georgia replaced New Jersey and Washington. There was less commonality with the top ten for total R&D among those states that performed the most federal intramural research. Only four states were found in both top-ten lists: Maryland, California, Texas and New Jersey.

Competition for resources is a fundamental explanation for the skewed distribution of R&D and science resources within the United States. States that lack such resources lag others in innovation, scholarship, graduate education, and overall economic growth as discussed by Kerr (1994) and Clark (1995).

Table 4.4. Selected Studies of Knowledge Spillovers

Spillover Source	Findings	Authors
Industry spillovers	Spillovers vary across industries; greater spillovers in knowledge-intensive industries	Jaffe (1989) Saxenian (1990) Acs, Audretsch, and Feldman (1992) Trajtenberg and Henderson (1993) Audretsch and Feldman (1996)
University spillovers	University spillovers more important to small firms than large firms	Link and Rees (1990) Audretsch and Feldman (1996)
Firm spillovers	Firm spillovers more important to large firms than small firms	Acs, Audretsch, and Feldman (1994) Feldman (1994) Edam, Levitas, and Martinez (1997)
City spillovers	Industrial diversity generates more spillovers than specialization; localized competition more than monopoly	Glaeser, Scheinkman, and Shleifer (1992) Almeida and Kogut (1997)
Individual spillovers	Spillovers shaped by role and mobility of knowledge workers	Audretsch and Stephan (1996) Prevezer (1997)

R&D-TO-PRODUCTIVITY GROWTH RELATIONSHIP

Since Solow's seminal article in 1957, in which he concluded that between 1909 and 1949 over 87 percent of the growth in the U.S. could not be explained by growth in capital and labor, researchers and policy makers have been looking for statistical correlates with this unexplained growth, which is commonly referred to as technological advancement or change. Hence, what these investigators have done is to posit that technological change, measured as total factor productivity growth, is causally related to increased investments in R&D. Total factor productivity growth is the best metric for improvements in economic efficiency.

Using manufacturing sector, industry, and firm-level data, researchers have examined the statistical association between R&D and total factor productivity growth. Their analytical models are not derived herein. Simply, these models correlate values in the growth of the shift parameter, A, from variants of equation (3) in Chapter 3, with measures of R&D expenditures in appropriate firms/industries. Mathematics aside, the extent of the correlation can be shown to be a measure of the marginal rate of return to R&D.

Researchers have estimated a relationship as:

(1) $$TFP = f(R\&D, \mathbf{X})$$

where **X** is a vector of other economic variables, and they have focused their interpretative attention on the coefficient of the R&D variable. In such econometric models R&D activity is measured in terms of R&D expenditures in total, by character of use, and by source of funding.

Table 4.5 summarizes the results of several key empirical studies. The key findings are as follows:

- The rate of return to privately-funded R&D is relatively large, ranging on average between 30 percent and 50 percent.
- The rate of return to privately-funded basic research is significantly greater than to privately-funded development, the differences being over 100 percent to basic research compared to about 15 percent to 20 percent for applied research plus development.
- The rate of return to federally-funded research performed in industry varies by character of use; the returns to federally-funded basic research performed in industry is over 100 percent, while federally-funded development has a negligible return on productivity growth.

Related studies have attempted to evaluate the social benefits, that is the spillover benefits to society, from industrial R&D. The rates of return to applied research and development described just above are for the most part private rates of return. More limited in number than the private rate of return studies, the findings from the social rate of return studies clearly indicate that the spillover benefits to society were somewhere in the 50 percent to 100 percent range.

Because of these findings, namely that the private and social rate of returns to R&D are relatively high, policy makers have remained focused on R&D investments in the private sector as a target variable for stimulating economic growth. The argument underlying such a focus is that, through incentives, firms will continue to invest in additional R&D projects and thus continue to stimulate economic growth and enhance standards of living through additional spillover effects.

Table 4.5. Selected Studies of the Relationship Between R&D and Productivity Growth

Findings	Authors
87.5 percent of the increase in aggregate output between 1909 and 1949 can be attributed to technical change.	Solow (1957)
R&D has positive impact on total factor productivity growth as does R&D embodied in purchased intermediate and capital goods.	Terleckyj (1974) Scherer (1982) Siegel (1997)
Rate of return to privately-financed basic research greater than for applied research or development.	Mansfield (1980) Link (1981) Griliches (1986) Lichtenberg and Siegel (1991)
Small direct impact of federally-financed R&D on total factor productivity growth.	Link (1981) Griliches (1986)

Most of the academic studies associated with this line of research were funded by the National Science Foundation during the late 1970s and early 1980s, motivated in large part by a slowdown in industrial productivity growth that began in the early 1970s[2] and increased in the late 1970s and early 1980s, (see Figure 3.2) and by the fact that U.S. industries were losing their competitive advantage in global markets.

[2] Researchers at that time were quick to point out that the slowdown in total factor productivity in the early 1970s was preceded by several years of flat and even declining R&D spending in the economy.

It is not surprising then that in the early 1980s, given the findings that the private and social rates of return to R&D were very high, that there were several important policy initiatives designed specifically to stimulate industrial R&D. We will discuss such initiatives in subsequent chapters.

5 THE ENTREPRENEURIAL PROCESS

From an economic standpoint, the concept of innovation can be traced at least as far back as the Physiocrats in France in the mid-1700s. Nicolas Baudeau (1910, p. 46) referred to a process guided by an active agent, which he called an entrepreneur, within a capitalistic system:

> Such is the goal of the grand productive enterprises: first to increase the harvest by two, three, four, ten times if possible; secondly to reduce the amount of labor employed and so reduce costs by a half, a third, a fourth, or a tenth, whatever possible.

Embedded in this conceptualization of entrepreneurship is the notion of an innovative process, one perhaps as simple as the perception of new technology adopted from others so as to increase agricultural yield, or one as refined as the actual development of a new technology to do the same. When the process is completed and when the innovation is put into use there will be a productivity gain and possibly even a substitution of capital for labor.

We have defined entrepreneurship as a process: an output is the promotion of one's own innovation or the adoption of another's innovation. The term entrepreneurship is commonly used to refer to a business person or even to a risk taker. Following Hébert and Link (1988), we use the term entrepreneur in a much broader sense; an entrepreneur is one who perceives an opportunity and has the ability to act upon it. Hence, entrepreneurship is a process that involves both *perception* and *action*. The perception of the opportunity may be influenced by changes in strategic directions or competitive markets, but perception of the opportunity is the fundamental first step. The consequent step is the ability to act on that perception. What defines the entrepreneur is the ability to move technology forward into innovation. The technology may be discovered or developed by others. The entrepreneur is able to recognize the commercial potential of the invention and organize the capital, talent, and other resources that turn an invention into a commercially viable innovation.

What are the requisite resources needed for action, which takes the perception of an opportunity forward to result in an innovation? One obvious answer is R&D because it not only provides a stock of knowledge to encourage perception but also the ability for the firm to foster action. However, firms that do not conduct R&D can still be entrepreneurial, as discussed above. In such firms, innovations are likely to be introduced rather than produced. Such firms act in an entrepreneurial manner by hiring creative individuals and providing them with an environment conducive for the blossoming of their talents.

Consider R&D-active firms. The R&D they conduct serves two general purposes. First, it provides the resource base from which the firm can respond to an opportunity with perceived strategic merit or technical opportunity that allows the firm to develop a commercial market. Second, those scientists involved in R&D are the internal resource that facilitates the firm's being able to make decisions regarding the technical merits of others' innovations and how effectively those innovations will interface with the existing technological environment of the firm. The firm may choose to purchase or license this technology or undertake a new R&D endeavor. In this latter sense, one related and very important role of R&D is to enhance the absorptive capacity of the firm (Cohen and Levinthal (1989).

Thus, the role of R&D as enhancing the absorptive capacity of the firm goes beyond simply assessing the technical merits of potentially purchasable technology. It allows the firm to interpret the extant technical literature; to interface when necessary with the research laboratories of others, in a research partnership relationship (discussed below) or to acquire technical explanations from, say, a federal laboratory or university laboratory; or simply to solve internal technical problems.

It is not surprising, then, that public policies aimed at enhancing the innovativeness of firms, especially in light of increasing global competition as was the case beginning in the late 1970s and early 1980s, have thus focused directly or indirectly on R&D.

Referring to Figure 3.3, within a single organization, in-house R&D leverages the entrepreneurial response to changes in either the strategic direction of the firm or its competitive market conditions. R&D is an appropriate policy target variable. However, although not shown in the figure, the firm's production process is characterized generally in terms of labor inputs and capital inputs. See equation (1) in Chapter 3. Capital inputs include new vintages of plant and equipment, and those new vintages embody the R&D that was conducted within the firm's that produced and supplies the purchased capital equipment.

Therefore public policy directed toward R&D affects a firm directly in terms of its internal entrepreneurial response, as in Figure 3.3, and indirectly in terms of the enhanced capital equipment that it purchases from other R&D-active entrepreneurial firms.

6 GOVERNMENT'S ROLE IN INNOVATION

The government should have an important role to play in fostering innovation, especially private-sector innovation. This intuitive conclusion logically follows from these facts:

- Innovation leads to technology.
- Technology is the prime driver of economic growth.
- In the absence of government intervention, firms will underinvest in technology, especially in basic research.
- Government has a responsibility to address this underinvestment or market failure by providing incentives to conduct additional R&D, which will enhance economic growth.

Such sequential reasoning to justify the role of government in innovation has dominated the history of public-sector involvement in the innovation process.

However, the economic underpinnings of government's role in innovation are more complex than might first appear. As described in this chapter, following Link and Scott (1998), the justification for the role of government in innovation, from an economic perspective, rests on a comparison of the efficiency of market resources with and without government intervention.

ECONOMIC RATIONALE FOR GOVERNMENT INVOLVEMENT

Even today, many policy makers and academics point to *Science—the Endless Frontier* to date the origins of U.S. science and technology policy. Bush did not articulate a science and technology policy, and he did not articulate an economic rationale for government's role. Rather, Bush like his predecessors and contemporaries simply assumed that the government had a role to play in the innovation process, and then he set out to describe that role (as opposed to offering a rationale for that role).

The first U.S. policy statement on science and technology was issued in 1990 by President George Bush, *U.S. Technology Policy*. As with any initial policy effort, this was an important general document. However, precedent aside, it too failed to articulate a rationale for government's intervention into the private sector's innovation process. Rather, like *Science—the Endless Frontier* and subsequent reports, it implicitly assumed that government had such a role, and it then set forth a rather general goal (1990, p. 2):

> The goal of U.S. technology policy is to make the best use of technology in achieving the national goals of improved quality of life for all Americans, continued economic growth, and national security.

President Clinton took a major step forward in his 1994 *Economic Report of the President* by first articulating principles about why the government has a role in innovation and in the overall technological process (p. 191):

> Technological progress fuels economic growth ... The Administration's technology initiatives aim to promote the domestic development and diffusion of growth- and productivity-enhancing technologies. They seek to correct market failures that would otherwise generate too little investment in R&D ... The goal of technology policy is not to substitute the government's judgment for that of private industry in deciding which potential "winners" to back. Rather the point is to correct [for] market failure ...

Although the 1994 *Economic Report of the President* did not expand on how to correct for market failure much less discuss appropriate policy mechanisms for doing so, it did for the first time posit an economic rationale for government involvement in the innovation process. Even more recently, the 1998 so-called Ehlers report, *Unlocking Our Future*, fails to articulate a clear rationale for government's involvement although it vastly departs in a positive way from the linear model of technology development proffered by Bush.

BARRIERS TO TECHNOLOGY AND MARKET FAILURE

Market failure refers to conditions under which the market, including those performing R&D and those adopting the R&D outputs of others, underinvests from society's perspective in any particular technology. Such underinvestment occurs because of conditions that prevent firms from fully realizing or appropriating the benefits expected from their investments.

Stated alternatively, firms underinvest in R&D when they determine, based on their expectations of post-innovation activity, that their private return is less than their private hurdle rate (minimum acceptable rate of return on their R&D investment). This is of public concern when the R&D investment not undertaken is socially desirable.

There are a number of factors that explain why a firm might perceive its private return to be less than its hurdle rate. These factors are what we call barriers to technology and they relate to technical and market risk, where risk is defined to measure the possibility that an actual outcome will deviate from an expected outcome.

Also, a firm may believe that it cannot appropriate a sufficient return on its R&D investment, even if it can overcome technical uncertainty. This may be due to a perceived inability to maintain proprietary control of the technology, thus enabling rivals to imitate their invention and reducing any resulting profitability.

Individually or in combination, the following factors contribute to why a firm may perceive a private rate of return as being less than its hurdle rate (Tassey 1999, National Science Board 2000):

- Because of high technical risk, the outcome of the R&D may not solve the technical problem adequately to meet perceived needs.
- Even if the R&D is technically successful, the market may not accept the technology because of competing alternatives or interoperability concerns.
- Even absent technical and market risk, it may be difficult to assign intellectual property rights to the technology, and it might be quickly imitated so that the innovator may not receive adequate return on the R&D investment.

These factors create barriers to investing in technology, which thus lead to market failure.

From an economic perspective, the role of government is to correct for these market failures in those instances where society will benefit from the technology. This latter situation occurs when the rate of return to society is greater than the social hurdle rate (minimum accepted rate of return for society from investments of resources with alternative social uses), as illustrated, following Tassey (1997) and Link and Scott (2001), in Figure 6.1.

The social rate of return is measured on the vertical axis along with society's hurdle rate on investments in R&D. The private rate of return is measured on the horizontal axis along with the private hurdle rate on investments in R&D.

A 45-degree line (long dashed line) is imposed on the figure. This is the point at which the social rate of return from an R&D investment equals the private rate of return from that same investment. The area above the 45-degree line and to the left of the private hurdle rate is the area of policy interest.

Figure 6.1. Gap Between Social and Private Rates of Return to R&D Projects

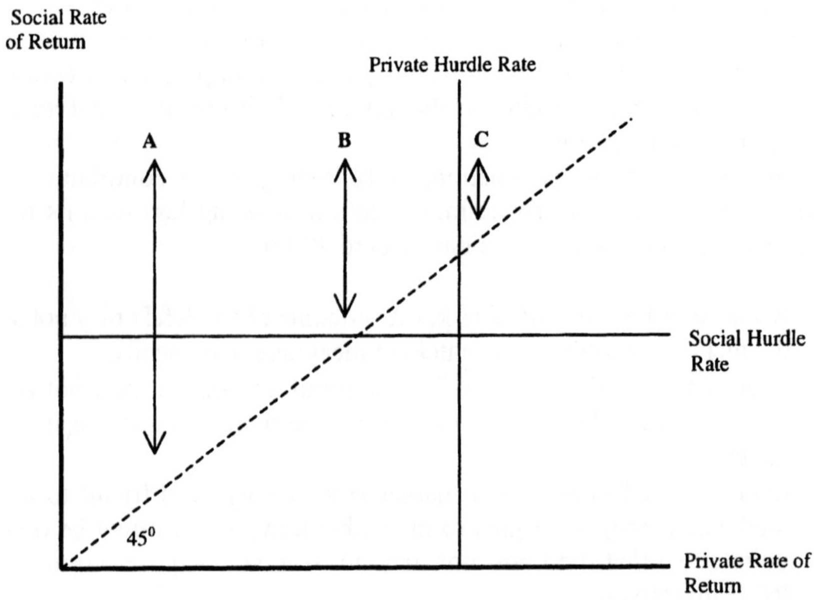

Three R&D projects are labeled in the figure. As drawn, the private rate of return exceeds the private hurdle rate for project C, and the social rate of return exceeds the social hurdle rate. The gap, measured vertically, between the social and private returns reflects the spillover benefits to society from the private R&D investment. However, the inability of the private sector to appropriate all benefits from its investment is not so great as to prevent the project from being adequately funded by the private firm.

In general, then, any R&D project that is expected to lead to a technology with a private rate of return to the right of the private hurdle rate and on or above the 45-degree line is in general not a candidate for public support, and hence government does not have an intervening role from an economic perspective.

Even in the presence of spillover benefits, project C will be funded by the private firm.

For projects A and B, the gap between the social and private returns is larger than in the case of project C; neither project A nor B will be adequately funded by the private firm as evidenced by the private rate of return being less than the private hurdle rate.

To address this market failure, the government has several policy mechanisms. Before examining specific policy mechanisms, a critical question is: Do firms, underinvest in R&D? Alternatively stated: Are there many projects like A and B that the private sector considers but rejects for some reason?

Private firms may not pursue promising technical opportunities for the following reasons:

- R&D scientific and technical frontiers are risky and the chances of failure are high.
- An individual firm may not have the capabilities required to develop the technology. Complex new technologies may require collaboration and information sharing; however, the cost of establishing research and development partnerships and making then work productively may provide disincentives to undertaking the project.
- Private incentives may not be sufficient to induce a firm to undertake the project in the face of difficulties in appropriating the resulting benefits (i.e., the resulting knowledge may flow to others who may benefit from the R&D without sharing the cost).

Government has at its disposal at least four policy mechanisms to reduce risk and market failure and thus overcome an underinvestment in R&D, where underinvestment refers to the situation where the private sector invests less in R&D than society would like it to invest (such as for project A and project B). These policy mechanisms include:

- Patent laws.
- Tax incentives.
- Improved environment for collaborative research.
- Subsidies to fund the research.

Each mechanism is discussed in the chapters that follow.

The U.S. patent system is more of an institution rather than a policy mechanism. Thus, patent laws, in a sense, characterize the overall innovative environment in which firms operate. In contrast, tax incentives, efforts to stimulate collaborative R&D, and direct and indirect government subsidies are in a more meaningful sense innovation policy levers.

7 THE PATENT SYSTEM

The history of the U.S. patent system dates to the authority given to Congress in the Constitution of the United States. Article I, section 8 states:

> Congress shall have power ... to promote the progress of science and useful arts, by securing for limited times to authors and inventors the exclusive right to their respective writings and discoveries.

Based on this authority, Congress initiated a number of patent laws beginning in 1790. The version of law that is now in effect was enacted on July 19, 1952 (to be effective January 1, 1953). It is Title 35 in the United States Code.

The Patent and Trademark Office issues patents for inventions. The patent term is 20 years, and it grants exclusive property rights to the inventor over that period of time. Patents are effective only within the United States and its territories and possessions.

U.S. law is clear about what can be patented. Any person who:

> invents or discovers any new or useful process, machine, manufacture, or composition of matter, or any new and useful improvement thereof, may obtain a patent.

It is important to note the word "useful," recalling that Franklin created the American Philosophical Society of Philadelphia in 1742 to promote "useful knowledge."

One cannot patent an idea. Three criteria must hold to be granted a patent: utility, novelty, and non-obviousness. Utility means that the invention must be useful; novelty means that the invention must be new and not merely a copy or repetition of another invention; and, non-obviousness—the most difficult criterion—means that the invention must neither be suggested by previous work nor totally anticipated given existing practices.

While the U.S. Code applies to patents granted in the United States, and in the territories and possessions of the United States, treaties have been promulgated to extend protection beyond national boundaries. The Paris Convention for the Protection of Intellectual Property of 1883 provided that each of the 140 signatory nations recognizes the patent rights of other countries. Subsequent treaties have extended such coverage and made filings in other countries more efficient.

Figure 7.1 illustrates what we call the economics of patenting from the perspective of the firm. The marginal private rate of return to R&D is measured on the vertical axis and the level of R&D spending is measured on the horizontal axis. The marginal private return schedules are downward sloping reflecting diminishing returns to R&D in any given time period, and for simplicity we assume the marginal private return schedule to be linear. Absent the patent system, the firm will choose to invest RD_0 in R&D. This is an optimal investment for the firm; it invests to the point where its marginal private cost of R&D (assumed to be constant for simplicity) equals its marginal private return. However, assume that the project society wants the firm to invest in requires $RD_1 > RD_0$.

Level RD_1 in Figure 7.1 is sufficient for the socially desirable project B in Figure 6.1 to be undertaken, but the firm does not have an incentive to invest in R&D at that level. The key point is that the existence of the patent, or more precisely, the expectation that the firm will be awarded a patent, can induce the firm to devote additional resources to R&D (to reach level RD_1.)

Receipt of monopoly power for 20 years through a patent increases the firm's marginal private return from its investments in R&D, thus shifting the marginal private return schedule to the right. The intersection of the new marginal private schedule and the marginal private cost schedule. The optimal investment for the firm is thus RD_1.

Several trends in patenting activity in the United States are noteworthy:

- In the early 1980s, the number of patents awarded to U.S. inventors began to decline and the number of U.S. patents awarded to foreign inventors began to rise, thus causing some policy makers to question the inventive environment in U.S. firms. This trend was yet another indicator that U.S. global competitiveness was declining.

- During the 1980s, the largest number of U.S. patents awarded to foreign inventors was granted to Japanese inventors. In fact, in 1995, over 20,000 patents were awarded to Japanese inventors, compared to about 7,000 for the next highest represented country, Germany.
- The share of total patents awarded to foreign inventors is low in the United States compared to other countries. It is highest in Italy and Canada and lowest in Japan and Russia.
- During the past decade, Japanese inventors have more international patents in three important technologies than inventors from any country, with the United States being second. These technologies are: robotics, genetic engineering, and advanced ceramics.
- Since the enactment of the Bayh-Dole Act of 1980 (see Chapter 10), which transferred ownership of intellectual property from federal agencies to universities, there has been a rapid rise in university patenting (Henderson, Jaffe, and Trajtenberg 1998).

Figure 7.1. Economics of Patenting: Increasing Marginal Private Return for the Firm

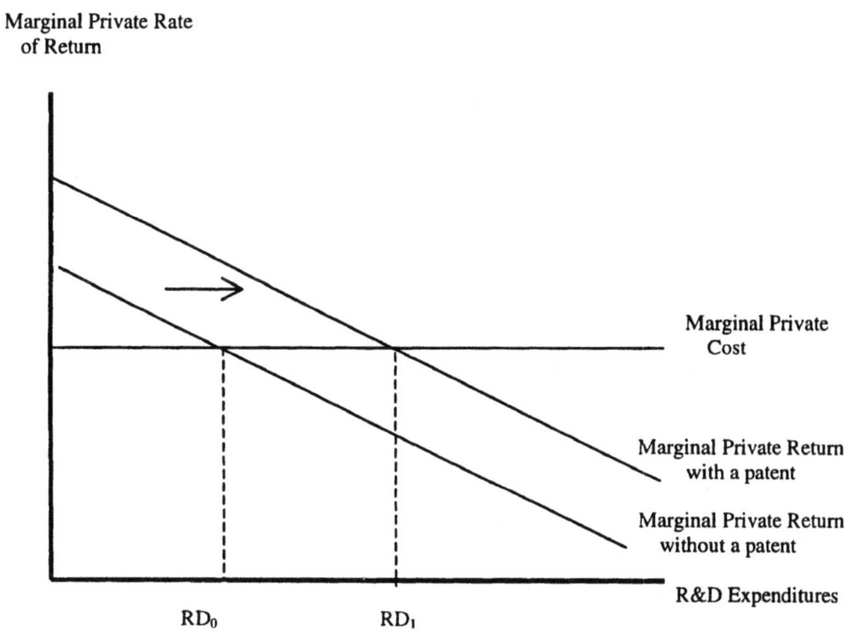

Researchers have investigated a number of economic issues related to patenting activity, and the economic role of patents in the innovation process. Some significant findings from this body of research are summarized in Table 7.1.

Table 7.1. Selected Literature on Patent Activity

Findings	Authors
Strong positive correlation between R&D expenditure (or employment) and patents	Scherer (1965) Schmookler (1966) Scherer (1983) Bound et al. (1984) Pakes and Griliches (1984) Hall, Griliches, and Hausman (1986) Acs and Audretsch (1989)
Positive correlation between patents and market value (stock market rate of return and Tobin's Q)	Griliches (1981) Hirschey (1982) Ben-Zion (1984) Pakes (1985) Cockburn and Griliches (1988) Austin (1993)
Value of patents is highly skewed, where value is determined by citations	Trajtenberg (1990a) Jaffe, Trajtenberg, and Henderson (1993) Henderson, Jaffe, and Trajtenberg (1998) Jaffe, Fogarty, and Banks (1998)
Citation-weighted measures of patents are more highly correlated with market value than unweighted measures.	Hall, Jaffe, and Trajtenberg (2000)

One key result from the above referenced literature is that the value of patents is highly skewed, where citations are used as an indicator of value. This result suggests that the use of counts of patents as an indicator of innovative output can be misleading if they are not properly deflated. Another critical finding is the existence of a strong positive correlation between R&D expenditure and patents. Finally, most studies report a positive correlation between patent activity and various measures of economic performance, including productivity, accounting profitability, Tobin's Q, and stock prices.

A recent paper by Hall, Jaffe, and Trajtenberg (2000) provides an important extension of this literature on estimation of the private returns (the returns that accrue to firms) to patenting by attempting to link citation-weighted patents to the market value of firms. Their preliminary results suggest that citation-weighted measures of patents are more highly correlated with market value than unweighted measures.

8 TAX INCENTIVES

*T*ax incentives are another mechanism that government uses to stimulate private sector R&D investment. Like any policy tool, tax incentives have advantages and disadvantages. Advantages include the following (Bozeman and Link 1984):

- Tax incentives entail less interference in the marketplace than do other mechanisms, thus affording private-sector recipients the ability to retain autonomy regarding the use of the incentives.
- Tax incentives require less paperwork than other programs.
- Tax incentives obviate the need to directly target individual firms in need of assistance.
- Tax incentives have the psychological advantage of achieving a favorable industry reaction.
- Tax incentives may be permanent and thus do not require annual budget review.
- Tax incentives have a high degree of political feasibility.

Some disadvantages of tax incentives are:

- Tax incentives may bring about unintended windfalls by rewarding firms for what they would have done in the absence of the incentive.
- Tax incentives often result in undesirable inequities.
- Tax incentives raid the federal treasury.
- Tax incentives frequently undermine public accountability.
- The effectiveness of tax incentives often varies over the business cycle.

THE ECONOMICS OF TAX CREDITS

Figure 8.1 illustrates the economics of a tax credit. The marginal rate of return is measured on the vertical axis and the level of R&D spending is measured on the horizontal axis. Both the marginal social return and the marginal private return schedules are downward sloping reflecting diminishing returns to R&D investments in a given time period. The social return schedule is drawn greater than the private return schedule for all levels of R&D because firms cannot appropriate all the benefits from conducting R&D; some of those benefits spillover to other firms in the current time period and in the post-innovation time period thus generating additional benefits to society. The marginal cost to the firm to undertake R&D is shown to be constant (horizontal).

Figure 8.1. Economics of a Tax Credit: Decreasing Marginal Private Cost for the Firm

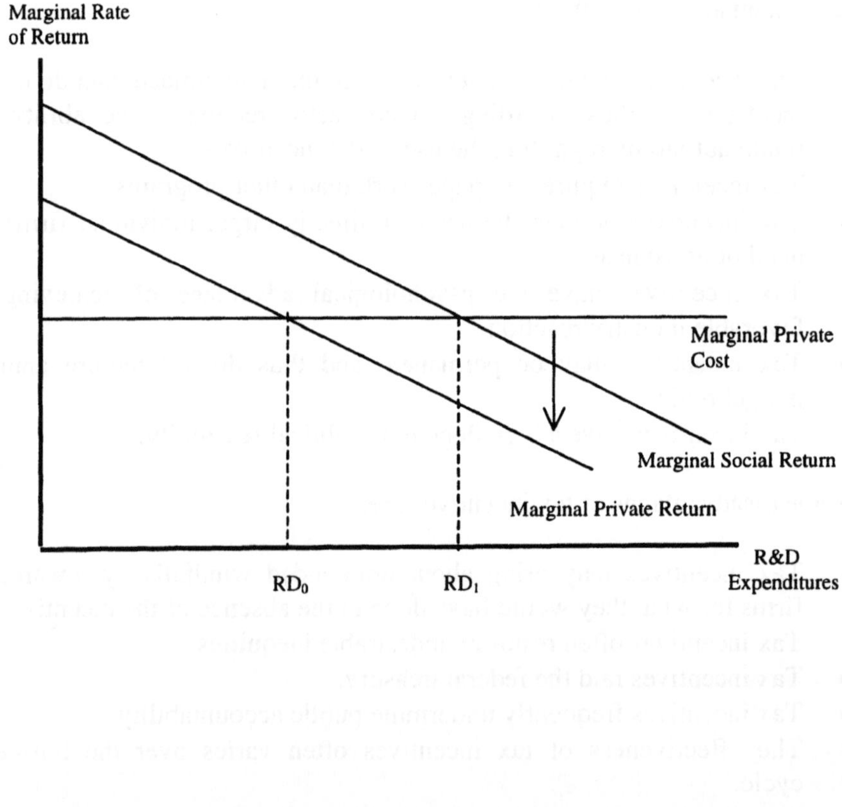

As drawn, the firm will equate the marginal private cost of conducting R&D with the marginal private return associated with this activity, and the firm will invest at the level RD_0. As in the patent example, assume that society wishes to push the firm to an investment level RD_1 to undertake a project like B in Figure 6.1. Society, given the firm's marginal cost schedule, would like the firm to invest in R&D to maximize social benefits. Hence, the optimal tax credit is one that provides an incentive to the firm to increase its R&D to point RD_1. Receipt of a tax credit can be thought of as a reduction in the marginal private cost of undertaking additional R&D, and the firm will re-equate its new marginal private cost with its marginal private return and invest at RD_1.

Unlike patents, the research and experimentation (R&E) credit simply increases the firm's private return on marginal R&D projects by reducing its marginal private cost to undertake such projects. Thus, tax incentives will increase the firm's level of R&D from RD_0 to RD_1 but will not alleviate the technical or market risk that characterizes the firm's portfolio of projects.

However, because R&D is not a homogeneous activity and because, as Mansfield (1980) and Link (1981) have shown, the research (R) portion of R&D has a greater impact on productivity growth and hence economic growth than does development (D), any uniform tax incentive that treats R&D as if it were a homogeneous activity will likely encourage more of the same mix of R&D. That may not necessarily be bad since economic studies suggest that the marginal private return from R&D is greater than the marginal cost of conducting R&D. However, a tax credit on research as opposed to development, while conceptually more desirable, could be cumbersome to administer.

THE R&E TAX CREDIT

The adoption of Section 174 of the Internal Revenue Code in 1954 codified and expanded tax laws pertaining to the R&D expenditures of firms. This provision permitted businesses to deduct fully R&E expenditures but not development or research application expenditures in the year incurred.

There is a slight distinction between R&D expenditures from a NSF-reporting perspective and R&E expenditures from a tax perspective. R&E expenditures are somewhat more narrowly defined to include all costs incident to development. R&E does not include ordinary testing or inspection of materials or products for quality control of those for efficiency studies, etc. R&E, in a sense, is the experimental portion of

R&D. That said, in practice it is often difficult to distinguish one category from the other.

Under Section 174, businesses are not allowed to expense R&E related equipment. Such equipment must be depreciated. However, the Economic Recovery and Tax Act of 1981 (ERTA) provided for a faster depreciation of R&E capital assets than other business capital assets.

ERTA also included a 25 percent tax credit for qualified R&E expenditures in excess of the average amount spent during the previous three taxable years or 50 percent of the current year's expenditures (the R&E base). The initial R&E tax credit had several limitations including the fact that it did not cover expenses related to the administration of R&D or to research conducted outside of the United States. The Tax Reform Act of 1986 modified these limitations, but reduced the marginal rate from 25 percent to 20 percent. Over the years the credit had been modified, primarily in terms of the definition of the R&E base, but the credit has never been made permanent. It has expired a number of times, only to be renewed retroactively. President George W. Bush has advocated for the credit to be made permanent.

In 1996, the Congressional Office of Technology Assessment released a report on the effectiveness of the R&E tax credit. The report concluded:

- There is not sufficient information available to conduct a complete benefit-to-cost analysis of the effectiveness of the R&E tax credit on the economy.
- The econometric studies that have been done to date conclude that the credit has been effective in the sense that for every dollar lost in federal revenue there is an increase of a dollar in private sector R&D spending. These studies conclude that the credit would be more effective if it were made permanent. Hall and van Reenen (2000) conclude from their review of the literature that the tax elasticity of R&D is about unity, meaning that a one percent increase in the credit will increase industry R&D by about one percent.
- The R&E tax credit represents a small fraction of federal R&D expenditures, about 2.6 percent of total federal R&D funding and about 6.4 percent of federal R&D for industry.

The R&E tax credit is not unique to the United States (Leyden and Link 1993). Japan's tax credit is marginal, and it was initiated in 1966. Canada also initiated a program in the 1960s, but their program is a flat tax program.

Some (Link and Bauer 1989) have proposed a tax credit for cooperative research. Economic theory concludes that firms that cooperate with each other in a research joint venture type of arrangement have the incentive to cooperate at the research end of the R&D spectrum rather than at the development end. Thus, a tax credit for cooperative research involvement is theoretically a viable alternative to the R&E tax credit and one that will potentially have a greater effect on research rather than development spending.

9 RESEARCH COLLABORATIONS

*R*esearch partnerships, meaning formal or informal collaboration among firms in the conduct of research, are an organizational form that may overcome elements of market failure by reducing technical risk to the R&D-conducting firms by enlarging the underlying knowledge base. Research partnerships may also limit market risk by helping to ensure that particular elements of the technology are standardized and thus interoperate in a system. As well, to the extent that collaboration reduces redundant research, there may be cost savings to each partner and reduced time to market.

Research partnerships are certainly not a new organizational form, but since the mid-1980s the federal government has adopted several policies to stimulate collaborative research. Before reviewing the related policies, we describe one research collaboration in detail as an illustration of the types of collaborations that are prevalent and that have had a major impact on research.

SEMICONDUCTOR RESEARCH CORPORATION

One of the first formal research collaborations in the United States was the Semiconductor Research Corporation (SRC). A brief history of the SRC will serve to illustrate that many research collaborations or partnerships are formed to address industry-wide technological issues, or at least issues that affect a sizeable segment of the industry. This brief history is also interesting because it illustrates, among other things, a purposeful entrepreneurial response to competitive market conditions.

In the late 1950s, an integrated circuit (IC) industry emerged in the United States. The fledgling industry took form in the 1960s and experienced rapid growth throughout the 1970s. In 1979, when Japanese companies captured 42 percent of the U.S. market for 16 kbit DRAMs (memory devices) and converted Japan's integrated circuit trade balance with the United States from a negative $122 million in 1979 to a positive $40 million in 1980, the U.S. industry became painfully aware that its dominance of the IC industry was being seriously challenged. It was clear to all in the industry that it was in their collective best interest to invest in an organizational structure that would strengthen the industry's position in the global semiconductor marketplace.

The Semiconductor Industry Association (SIA) was formed in 1977 to collect and assemble reliable information on the industry and to develop mechanisms for addressing industry issues with the federal government. In a presentation at an SIA Board Meeting in June 1981, Erich Bloch of IBM described to the industry the nature of the growing competition with Japan and proposed the creation of a "semi-conductor research cooperative" to assure continued U.S. technology leadership. This event witnessed the birth of the SRC. In December 1981, Robert Noyce, then SIA chairman and vice-chairman of Intel, announced the establishment of the SRC for the purpose of stimulating joint research in advanced semiconductor technology by industry and U.S. universities and to reverse the declining trend in semiconductor research investments. The SRC was formally incorporated in February 1982 with a stated purpose to:[1]

- Provide clearer view of technology needs.
- Fund research to address technology needs.
- Focus attention on competition.
- Reduce research redundancy.

Policy makers soon noticed the virtues of cooperative research in part because such organizational structures had worked well in Japan and in part because the organizational success of the SRC demonstrated that cooperation among competitive firms at the fundamental research level was feasible.

---

[1] The eleven founding members were Advanced Micro Devices, Control Data Corporation, Digital Equipment Corporation, General Instrument, Honeywell, Hewlett-Packard, IBM, Intel, Monolithic Memories, Motorola, National Semiconductor, and Silicon Systems.

PUBLIC POLICY TOWARD RESEARCH COLLABORATIONS

To place the activities surrounding the SRC's formation in a broader context, recall that in the late 1970s and early 1980s there was growing concern regarding the pervasive slowdown in productivity growth throughout the U.S. industrial sector. More specifically, policy makers were troubled by the declining global market shares of leading American companies, especially firms in the semiconductor industry.[2]

As noted in a November 18, 1983 House report about the proposed Research and Development Joint Ventures Act of 1983:

> A number of indicators strongly suggest that the position of world technology leadership once firmly held by the United States is declining. The United States, only a decade ago, with only five percent of the world's population was generating about 75 percent of the world's technology. Now, the U.S. share has declined to about 50 percent and in another ten years, without fundamental changes in our Nation's technological policy ... the past trend would suggest that it may be down to only 30 percent. [In hearings,] many distinguished scientific and industry panels had recommended the need for some relaxation of current antitrust laws to encourage the formation of R&D joint ventures. ... The encouragement and fostering of joint research and development ventures are needed responses to the problem of declining U.S. productivity and international competitiveness. According to the testimony received during the Committee hearings, this legislation will provide for a significant increase in the efficiency associated with firms doing similar research and development and will also provide for more effective use of scarce technically trained personnel in the United States.

---

[2] The declining U.S. position in the semiconductor industry was well known and in other industries there was widespread concern although the empirical evidence about the competitive position of the United States in international markets was incomplete. However, when the U.S. Department of Commerce (1990) released its 1990 report on emerging technologies, it was apparent to all that the concerns expressed in the early 1980s were quite valid.

In an April 6, 1984 House report on competing legislation, the Joint Research and Development Act of 1984, the supposed benefits—and recall that at this time it was still too soon for there to be visible benefits coming from the SRC's activities on behalf of the IC industry—of joint research and development were for the first time clearly articulated:

> Joint research and development, as our foreign competitors have learned, can be procompetitive. It can reduce duplication, promote the efficient use of scarce technical personnel, and help to achieve desirable economies of scale. ... [W]e must ensure to our U.S. industries the same economic opportunities as our competitors, to engage in joint research and development, if we are to compete in the world market and retain jobs in this country.

The National Cooperative Research Act (NCRA) of 1984, after additional revisions in the initiating legislation, was passed on October 11, 1984:

> to promote research and development, encourage innovation, stimulate trade, and make necessary and appropriate modifications in the operation of the antitrust laws.

The NCRA created a registration process, later expanded by the National Cooperative Research and Production Act (NCRPA) of 1993, under which research joint ventures (RJVs) can disclose their research intentions to the Department of Justice. RJVs gain two significant benefits from such voluntary filings: if subjected to criminal or civil action they are evaluated under a rule of reason that determines whether the venture improves social welfare; and if found to fail a rule-of-reason analysis they are subject to actual rather than treble damages.

One of the more notable RJVs formed and made public through the NCRA disclosure process was SEMATECH (SEmiconductor MAnufacturing TECHnology). It was established in 1987 as a not-for-profit research consortium with an original mission to provide a pilot manufacturing facility where member companies could improve their semiconductor manufacturing process technology. Its establishment came after the Defense Science Board recommended direct government subsidy to the industry in a 1986 report commissioned by the Department of Defense (therefore SEMATECH is discussed below under the broader heading of a public/private technology partnership). It was thought that SEMATECH would be the U.S. semiconductor industry's/U.S. government's response to the Japanese government's targeting of their semiconductor industry for global domination. Since its inception, SEMATECH's stated mission has evolved and become more general. The consortium currently defines its mission as solving the technical challenges presented in order to sustain a leadership position for the United States in the global semiconductor industry.

TRENDS IN RJVs

To date, there have been over 800 formal RJVs filed under the NCRA. Certainly, this number is a lower bound on the total number of research partnerships in the United States, even since 1984. Not all RJVs are as publicly visible as SEMATECH. Most are quite small, with only two or three members, and others are quite large with hundreds of members. The average size of one of these joint ventures is about 14 members.

As an illustration of the research activity that can successfully occur through a small, less visible research partnership, consider the Southwest Research Institute Clean Heavy Diesel Engine II joint venture, noticed in the *Federal Register* in early-1996. The eleven member companies, from six countries including the United States, joined together to solve a common set of technical problems. Diesel engine manufacturers were having difficulties, on their own, meeting desired emission control levels. The eleven companies were coordinated by Southwest Research Institute, an independent, non-profit contract research organization in San Antonio, Texas, to collaborate on the reduction of exhaust emissions. The joint research was successful, and each member company took with it fundamental process technology to use in their individual manufacturing facilities to meet desired emission control levels. The joint venture was formally disbanded in mid-1999.

While informal cooperation in research may have been prevalent in the United States for decades, formal RJV relationships are new and it will take longer than a decade and a half to detect meaningful trends (Brod and Link 2001).

Although research partnerships as formal entities are relatively new to the technology strategy arena, the literature concludes that there are both benefits and costs to members of the venture. Following Hagedoorn, Link, and Vonortas (2000), the benefits include:

- The opportunity for participants to capture knowledge spillovers from other members.
- Reduced research costs due to a reduction in duplicative research.
- Faster commercialization since the fundamental research stage is shortened.
- The opportunity to develop industry-wide competitive vision.

The costs include:

- A lack of appropriability since research results are shared among the participants.
- Managerial tension, in some cases, as participants learn to trust each other and to work together.

Research partnerships are correctly viewed as a complementary source of technical knowledge and technical efficiency for the firm. Thus, firms that participate in a research partnership enhance their own R&D process through interactions.

UNIVERSITIES AS RESEARCH PARTNERS

Especially noticeable in the RJVs filed with the Department of Justice is the presence of universities as research partners. Over the past 15 years, the number of RJVs with at least one university partner has increased. On average, about 15 percent of RJVs have, as of 2000, at least one university research partner, and of these over 90 percent are U.S. universities. Those RJVs with universities as research partners have on average five university partners.

The literature on universities as research partners is sparse. However, some stylized conclusions can be drawn from the limited investigations (Hall, Link, Scott 2000, 2001):

- Firms that interact with universities generally have greater R&D productivity and greater patenting activity.
- One key motive for firms to maintain joint research relationships with universities is to have access to key university personnel—faculty as well as students as potential employees.

Many commentators predict that university participation in such collaboration will increase in the future. According to the Council on Competitiveness (1996, pp. 3-4):

> Over the next several years, participation in the U.S. R&D enterprise will have to continue experimenting with different types of partnerships to respond to the economic constraints, competitive pressures and technological demands that are forcing adjustment across the board. ... [and in response] industry is increasingly relying on partnerships with universities, while the focus of these partnerships is shifting progressively toward involvement in shorter-term research.

And (Council on Competitiveness 1996, p. 11):

> For universities, cutbacks in defense spending have resulted in a *de facto* reallocation of funding away from the physical sciences and engineering and shifted the focus of defense research away from the frontiers of knowledge [e.g., basic science] to more applied efforts. ... Although defense spending is clearly not the only viable mechanism to support frontier research and advanced technology, the United States has yet to find an alternative innovation paradigm to replace it.

Given this spending trend, and the increasing ease of global technology transfer, it is conceivable, at least according to the Council, that the United States may lose its technological leadership in some important areas such as health and advanced materials, since innovation in these fields is closely linked to improvements in basic science.

There is some indication that scholars are beginning to think more deeply and more broadly about the social, economic, and technological consequences of university involvement in private sector research partnerships (Siegel et al. 2001). This thinking reflects some major concerns about the impact of these relationships on the research university's mission to conduct basic research. Unfortunately, there is a void of information that can be studied by researchers to examine the ramifications of this trend from a wide variety of disciplinary perspectives.

It is likely that the increasing trend toward university private-sector research partnerships will continue. A 1993 national survey of U.S. biology, chemistry, and physics faculty members revealed that many academic scientists desired more of such involvement. An earlier survey of engineering faculty members reached the same conclusion. However, one of the authors (Morgan 1998, p. 169) of the surveys was quick to point out one area of major concern:

> a diminution of the role of the university as an independent voice to help look out for the broader societal good and to guard against industrial as well as other excesses. An independent science, engineering and public policy role is essential to ensure an adequate supply of well educated scientists and engineers prepared to work with the public sector in public interest groups. Having industry assume a more central role as customer and client for university-based scientific and engineering research, while in some way a natural and desirable step, needs to be balanced against the need for independence, oversight and service to society and the larger public good.

GOVERNMENT LABORATORIES AS A RESEARCH PARTNER

The federal government also enters directly into research partnerships with firms through the federal laboratory system. This relationship can take various forms ranging from informal relationships whereby a firm(s) interacts with a federal laboratory scientist, or more formal relationships whereby a firm(s) utilizes federal laboratory facilities and is jointly involved with the laboratory scientists in the research. Or, the relationship can be nothing more than an informational transfer whereby the firm utilizes public information that was generated within a government agency.

While very few studies have systematically looked at the economics of federal laboratories as research partners, two generalizations can be made (Leyden and Link 1999):

- Federal laboratories are generally associated with research joint ventures that are large in terms of other member companies.
- One key advantage to partnering with a federal laboratory is access to specialized technical equipment.

In response to the Government Performance and Results Act of 1993 (discussed in Chapter 14), federal laboratories, because of their research nature, must document performance outputs and outcomes. As information on these outputs and outcomes is disseminated, more will be known about the role of federal laboratories in leveraging private-sector R&D and the spillover benefits associated with the fundamental knowledge that they generate.

10 PUBLIC/PRIVATE PARTNERSHIPS

*O*ne can trace the origins of public/private partnerships—federal grants assistance technology partnerships—in the United States at least as far back as the Lincoln Administration. As noted in Link (1999), the Morrill Act of 1862 established what was known as the land-grant college system. In effect, this legislation created a partnership between the federal and state governments to cooperate with the private sector in technology development. The Act charged states to develop colleges to offer curricula in agriculture and mechanical arts. Then in 1887, the Hatch Act provided resources for a system of state agriculture experiment stations that would be under the auspices of land-grant colleges and universities. A partnership among the various levels of government was established by the Smith-Level Act of 1914. The Cooperative Agriculture Extension service was charged to deliver the practical benefits of research to citizens through an extension service.

According to Carr (1995, p. 11):

> Until the end of the 1970s, the philosophy behind the dissemination of federally-funded research was that if the public paid for the research, the resulting intellectual property should be made equally available to all interested parties.

Because the 1970s and 1980s witnessed many foreign competitors beginning to successfully challenge the long-standing dominance of the United States both in world and domestic markets, it became clear to public and private policymakers that a change in the philosophy of federal R&D support was needed. The Office of Technology Policy (1996) reflected on the motivation for this change in policy mindset as:

- Global competitors were better able to appropriate the output of U.S. basic and mission-oriented research as their technical sophistication grew.
- Traditional public-sector mechanisms of technology development, transfer, and development took too long in an era of accelerating private sector development.
- U.S. federal R&D represented a declining share of the world R&D as globally-competitive nations increased their public funding; hence the marginal benefits to industry from additional public moneys (allocated in the same historical manner) declined.

Stated alternatively, but maintaining the same general theme, Link and Tassey (1987, pp. 4, 131) reflected on the changing competitive environment of U.S. industry in the early 1980s:

> there is a new order of competition in the world. An inescapable element of the competition is technology ... With the advent of technology-based economies [throughout the world], the increase in the number of world competitors has been greater than the increase in the size of the world market. What has resulted from this is a significant shortening of technology life cycles ... As such, effective long-run competitive strategies will have to deal explicitly with technology ... [C]ompetitive survival will depend on technology-based strategies. These strategies will have to evolve from new philosophies about interdependence ... The importance of interdependence arises from the need of a domestic industry to rapidly and efficiently develop complex technological elements from which specific applications (innovations) are drawn for competitive activity ... [Accordingly,] government must expand and adapt its role ... with industry for more effective joint planning in research.

Beginning with legislation in the 1980s, as summarized in Table 10.1, a new era in federal technology policy began. This era was based on the belief that the global competitiveness of U.S. firms can be enhanced through legislation to bolster the commercial impact of federal R&D investments. As such, using the terminology of the Office of Technology Policy (1996, p. 26), "a new paradigm for public/private technology partnerships emerged." This new paradigm viewed industry as a partner in the formation and execution of technology programs rather than a passive recipient of the output from federal research.

Several public/private partnerships are overviewed below from an institutional perspective. SEMATECH is discussed in moderate detail because it was one of the earliest public/private partnerships and for years was heralded as the model organizational form for other public/private partnerships to follow. The Small Business Innovation Research Program (SBIR) and the Advanced Technology Program (ATP) are also singled out for discussion because these are two of the programs that have conducted in-depth program evaluations and hence more is known about their economic impacts than about those associated with other partnerships.

SEMATECH

In 1986 when the Semiconductor Industry Association (SIA) and the Semiconductor Research Corporation (SRC) began to explore the possibility of joint industry/government cooperation, the U.S. semiconductor industry was not in a favorable economic position. During 1986, Japan overtook the United States for the first time in terms of their share of the world semiconductor market. Japan had about 45 percent of the world market compared to about 42 percent for the United States. The U.S. semiconductor industry expected Japan's share to grow at the expense of that of the United States.

In January 1987, President Reagan recommended $50 million in matching federal funding for R&D related to semiconductor manufacturing, and this was to be part of the Department of Defense's 1988 budget. Soon thereafter, the SIA approved the formation of SEMATECH and the construction of a world-class research facility.[1] In September 1987, Congress authorized $100 million in matching funding for SEMATECH.

[1] The thirteen charter members of SEMATECH were: Advanced Micro Devices, AT&T, Digital Equipment Corporation, Harris Corporation, Hewlett-Packard Company, IBM Corporation, Intel Corporation, LSI Logic Corporation, Micron Technology, Inc., Motorola, Inc., National Semiconductor Corporation, Rockwell International Corporation, and Texas Instruments, Inc.

Table 10.1. Selected Public/Private Technology Partnership Legislation

Enabling Legislation	Characteristics of the Program
Stevenson-Wylder Technology Innovation Act of 1980	Act predicated on the premise that federal laboratories embody industrially-useful technology. Federal laboratory mandated to establish an Office of Research and Technology Application to facilitate transfer of public technology to the private sector.
University and Small Business Patent Procedure Act of 1980	Known as Bayh-Dole Act. Reformed federal patent policy by providing increased incentives for diffusion of federally-funded innovation results. Universities, non-profit organizations, and small businesses permitted to obtain titles to innovations developed with governmental funds; federal agencies to grant exclusive licenses to their technology to industry.
Small Business Innovation Development Act of 1982	Act required federal agencies to provide special funds to support small business R&D that complemented the agency's mission. Programs called Small Business Innovation Research (SBIR) programs. Act reauthorized in 1992.
National Cooperative Research Act of 1984	NCRA encouraged formation of joint research venture among U.S. firms. Amended by the National Cooperative Research and Production Act of 1993, thereby expanding antitrust protection to joint production ventures.
Trademark Clarification Act of 1984	Act set forth new licensing and royalty regulations to take technology from federally-funded facilities into the private sector. Specifically permitted government-owned, contractor-operated (GOCO) laboratories to make decisions regarding which patents to license to the private sector, and contractors could receive royalties on such patents.
Federal Technology Transfer Act of 1986	Act amended by the Stevenson-Wylder Act. Made technology transfer an explicit responsibility of all federal laboratory scientists and engineers. Authorized cooperative research and development agreements (CRADAs). Amended by the National Competitiveness Technology Transfer Act of 1989 to include contractor operated laboratories.
Omnibus Trade and Competitiveness Act of 1988	Act established Advanced Technology Program (ATP) and Manufacturing Extension Partnership (MEP) within the re-named National Institute of Standards and Technology (NIST).
Defense Conversion, Reinvestment, and Transition Assistance Act of 1992	Act created infrastructure for dual-use partnerships. Through Technology Reinvestment Project partnerships, Department of Defense given the ability to leverage the potential advantages of advanced commercial technologies to meet departmental needs.

SEMATECH and its members have a mission to:

> create a shared competitive advantage by working together to achieve and strengthen manufacturing technology leadership.

This shared vision is accomplished by joint sponsorship of leading edge technology development in equipment supplier companies. As these companies become world class manufacturers, so will the members of SEMATECH.

By 1988, Japan's world market share reached over 50 percent, and that of the United States fell to about 37 percent. The U.S. share declined again in 1989 and then it began to increase at the expense of that of Japan. Early in 1992, the United States was again at parity with Japan at about 42 percent, and stayed slightly ahead of Japan until 1995 when the gap began to widen.

The mid-1990s saw increasing cooperation between U.S. and Japanese semiconductor companies, and in fact, in 1998 International SEMATECH began operations with Hyundai (Japan) and Philips (Amsterdam) as important members. Also, beginning in 1998, all funding came only from member companies.

SMALL BUSINESS INNOVATION RESEARCH PROGRAM

The Small Business Innovation Research (SBIR) program began at the National Science Foundation (NSF) in 1977 (Tibbetts 1999). At that time the goal of the program was to encourage small businesses, long believed to be engines of innovation in the U.S. economy, to participate in NSF-sponsored research, especially research that had commercial potential. Because of the early success of the program at NSF, Congress passed the Small Business Innovation Development Act of 1982. The Act required all government departments and agencies with external research programs of greater than $100 billion to establish their own SBIR programs and to set aside funds equal to 0.2 percent of the external research budget. As a set aside program, the SBIR program redirects existing R&D rather than appropriating new monies for R&D. Currently, agencies must allocate 2.5 percent of the external research budget to SBIR.

The 1982 Act states that the objectives of the program are:

(1) to stimulate technological innovation
(2) to use small business to meet Federal research and development needs
(3) to foster and encourage participation by minority and disadvantaged persons in technological innovation
(4) to increase private sector commercialization of innovations derived from federal research and development.

The Act was reauthorized in 1992 with the same objectives.

SBIR awards are of three types. Phase I awards are small, generally less than $100,000. The purpose of these awards is to assist firms to assess the feasibility of the research they propose to undertake for the agency in response to the agency's objectives. Phase II awards can range up to $750,000. These awards are for the firm to undertake and complete its proposed research, hopefully leading to a commercializable product or process.

The Department of Defense's (DoD's) SBIR program has been evaluated in some detail (Wessner 2000). It can be concluded that DoD's SBIR program in encouraging commercialization from research that would not have been undertaken without SBIR support. And moreover, the structure of DoD's SBIR program is overcoming reasons for market failure that previously have caused small firms to underinvest in R&D. In fact, Audretsch, Link, and Scott (forthcoming) found that SBIR R&D does lead to commercialization, and that the net social benefits associated with the program's sponsored research are substantial.

ADVANCED TECHNOLOGY PROGRAM

The Advanced Technology Program (ATP) was established within the National Institute of Standards and Technology (NIST) through the Omnibus Trade and Competitiveness Act of 1988, and modified by the American Technology Preeminence Act of 1991. The goals of the ATP, as stated in its enabling legislation, are to assist U.S. businesses in creating and applying the generic technology and research results necessary to:

(1) commercialize significant new scientific discoveries and technologies rapidly
(2) refine manufacturing technologies.

These same goals were restated in the *Federal Register* on July 24, 1990:

> The ATP . . . will assist U.S. businesses to improve their competitive position and promote U.S. economic growth by accelerating the development of a variety of pre-competitive generic technologies by means of grants and cooperative agreements.

The ATP received its first appropriation from Congress in FY 1990. The program funds research, not product development. Commercialization of the technology resulting from a project might overlap the research effort at a nascent level, but generally full translation of the technology into products and processes may take a number of additional years. ATP, through cost sharing with industry, invests in risky technologies that have the potential for spillover benefits to the economy.

Appropriations to ATP increased from $10 million in 1990 to a peak of $341 million in 1995. Funding decreased in 1996 to $221 million, and it has averaged about $200 million per year until 2000 when it fell to just under $150 million. To date, ATP has funded through competitive processes approximately 450 research projects.

Much like the case of DoD's SBIR program, ATP has provided incentives to firms to undertake research that would not otherwise have been pursued—like projects A or B in Figure 6.1 (Link and Scott 2001).

11 INFRASTRUCTURE TECHNOLOGY

Technology infrastructure is a term that refers to the technological environment in which firms innovate. Two critical aspects of the technological environment of firms are patent laws (previously discussed) and the federal laboratory system (publicly funded). The federal laboratory system's output is infrastructure technology.

FEDERAL LABORATORY SYSTEM

As shown above in Table 4.1, the federal government allocated $65.9 billion to R&D in 1999, of which $17.4 billion was performed by the federal government. Most of this research occurred in the more than 700 federally-funded R&D laboratories in the country; some of these laboratories are government owned but contractor operated (GOCO) and others are government owned and government operated (GOGO). The Department of Defense's laboratories account for almost one-half of this intramural research, and the Department of Health and Human Services accounts for about 15 percent.

Table 11.1 provides some summary information about a few of the laboratories in the federal laboratory system.

The governmental agency that provides a significant amount of direct infrastructure technology to the industrial economy is the National Institute of Standards and Technology (NIST) within the Department of Commerce. Emphasis is placed throughout this chapter on direct technology infrastructure because a significant amount of indirect technology infrastructure comes from all departments and agencies as technological knowledge spillover and is transferred to industry in one form or another.

NIST is discussed herein in more detail than any other federal laboratory for two reasons: one, its research mission is varied by field of science, as opposed to being focused on one major scientific area such as energy; and two, the Program Office within NIST has a long history of evaluation of the outputs from its research programs, and as such its contribution to economic growth is more readily documented.

Table 11.1. Overview of Selected National Laboratories

Laboratory	Year Established	Research	Ownership
Ames Laboratory	1942	Energy	Department of Energy operated by Iowa State University
Brookhaven National Laboratory	1947	Energy	Department of Energy operated by SUNY Stony Brook
Los Alamos National Laboratory	1943	National security and energy	Department of Energy
National Institute of Standards and Technology	1901	Standards	Department of Commerce
Rome Laboratory	1951	Communications	Department of Defense

Source: Crow and Bozeman (1998)

NATIONAL INSTITUTE OF STANDARDS AND TECHNOLOGY

A standard is a prescribed set of rules, conditions, or requirements concerning:

- Definitions of terms.
- Classification of components.
- Specification of materials, their performance, and their operations.
- Delineation of procedures.

- Measurement of quantity and quality in describing materials, products, systems, services, or practices.

To understand the current activities that take place at NIST, its public good mission must be placed in an historical perspective. The concept of the government's involvement in standards traces to the Articles of Confederation signed on July 9, 1778. In Article 9, § 4:

> The United States, in Congress assembled, shall also have the sole and exclusive right and power of regulating the alloy and value of coin struck by their own authority, or by that of the respective States; fixing the standard of weights and measures throughout the United States ...

This responsibility was reiterated in Article 1, § 8 of the Constitution of the United States:

> The Congress shall have power ... To coin money, regulate the value thereof, and of foreign coin, and fix the standard of weights and measures ...

On July 20, 1866, Congress and President Andrew Johnson authorized the use of the metric system in the United States. This was formalized in the Act of 28 July 1866—An Act to Authorize the Use of the Metric System of Weights and Measures:

> *Be it enacted* ..., That from and after the passage of this act it shall be lawful throughout the United States of America to employ the weights and measures of the metric system; and no contract or dealing, or pleading in any court, shall be deemed invalid or liable to objection because the weights or measures expressed or referred to therein are weights and measures of the metric system.

As background to this Act, the origins of the metric system can be traced to the research of Gabriel Mouton, a French vicar, in the late 1600s. His standard unit was based on the length of an arc of 1 minute of a great circle of the earth. Given the controversy of the day over this measurement, the National Assembly of France decreed on May 8, 1790 that the French Academy of Sciences along with the Royal Society of London deduced an invariable standard for all the measures and all the weights.

Within a year, a standardized measurement plan was adopted based on terrestrial arcs, and the term mètre (meter), from the Greek *metron* meaning to measure, was assigned by the Academy of Sciences (Link 1996a).

Because of the growing use of the metric system in scientific work rather than commercial activity, the French government held an international conference in 1872, which included the participation of the United States, to settle on procedures for the preparation of prototype metric standards. Then, on May 20, 1875, the United States participated in the Convention of the Meter in Paris and was one of the eighteen signatory nations to the Treaty of the Meter.

In a Joint Resolution before Congress on March 3, 1881, it was resolved that:

> The Secretary of the Treasury be, and he is hereby directed to cause a complete set of all the weights and measures adopted as standards to be delivered to the governor of each State in the Union, for the use of agricultural colleges in the States, respectively, which have received a grant of lands from the United States, and also one set of the same for the use of the Smithsonian Institution.

Then, the Act of 11 July 1890 gave authority to the Office of Construction of Standard Weights and Measures (or Office of Standard Weights and Measures), which had been established in 1836 within the Treasury's Coast and Geodetic Survey:

> For construction and verification of standard weights and measures, including metric standards, for the custom-houses, and other offices of the United States, and for the several States ...

The Act of 12 July 1894 established standard units of electrical measure:

> *Be it enacted* ..., That from and after the passage of this Act the legal units of electrical measure in the United States shall be as follows: ... That it shall be the duty of the National Academy of Sciences [established in 1863] to prescribe and publish, as soon as possible after the passage of this Act, such specifications of detail as shall be

> necessary for the practical application of the definitions of the ampere and volt hereinbefore given, and such specifications shall be the standard specifications herein mentioned.

Following from a long history of our nation's leaders calling for uniformity in science, traceable at least to the several formal proposals for a Department of Science in the early 1880s, and coupled with the growing inability of the Office of Weights and Measures to handle the explosion of arbitrary standards in all aspects of federal and state activity, it was inevitable that a standards laboratory would need to be established. The political force for this laboratory came in 1900 through Lyman Gage, then Secretary of the Treasury under President William McKinley.

Gage's original plan was for the Office of Standard Weights and Measures to be recognized as a separate agency called the National Standardizing Bureau. This Bureau would maintain custody of standards, compare standards, construct standards, test standards, and resolve problems in connection with standards. Although Congress at that time wrestled with the level of funding for such a laboratory, the importance of the laboratory was not debated.

Finally, the Act of 3 March 1901, also known as the Organic Act, established the National Bureau of Standards within the Department of the Treasury, where the Office of Standard Weights and Measures was administratively located:

> *Be it enacted by the Senate and House of Representatives of the United States of America in Congress assembled,* That the Office of Standard Weights and Measures shall hereafter be known as the National Bureau of Standards ... That the functions of the bureau shall consist in the custody of the standards; the comparison of the standards used in scientific investigations, engineering, manufacturing, commerce, and educational institutions with the standards adopted or recognized by the Government; the construction, when necessary, of standards, their multiples and subdivisions; the testing and calibration of standard measuring apparatus; the solution of problems which arise in connection with standards; the determination of physical constants and the properties of materials, when such data are of great importance to scientific or manufacturing interests and are not to be obtained of sufficient accuracy elsewhere.

The Act of 14 February 1903, established the Department of Commerce and Labor, and in that Act it was stated that the National Bureau of Standards be moved from the Department of the Treasury to the Department of Commerce and Labor.

Then, in 1913, when the Department of Labor was established as a separate entity, the Bureau was formally housed in the Department of Commerce.

In the post World War I years, the Bureau's research focused on assisting in the growth of industry. Research was conducted on ways to increase the operating efficiency of automobile and aircraft engines, electrical batteries, and gas appliances. Also, work was begun on improving methods for measuring electrical losses in response to public utility needs. This latter research was not independent of international efforts to establish electrical standards similar to those established over 50 years earlier for weights and measures.

After World War II, significant attention and resources were given to the activities of the Bureau. In particular, the Act of 21 July 1950 established standards for electrical and photometric measurements:

> *Be it enacted by the Senate and House of Representatives of the United States of America in Congress assembled,* That from and after the date this Act is approved, the legal units of electrical and photometric measurements in the United States of America shall be those defined and established as provided in the following sections. ... The unit of electrical resistance shall be the ohm ... The unit of electrical current shall be the ampere ... The unit of electromotive force and of electrical potential shall be the volt ... The unit of electrical quantity shall be the coulomb ... The unit of electrical capacity shall be the farad ... The unit of electrical inductance shall be the henry ... The unit of power shall be the watt ... The units of energy shall be the (a) joule ... and (b) the kilowatt-hour ...The unit of intensity shall be the candle ... The unit of flux light shall be the lumen ... It shall be the duty of the Secretary of Commerce to establish the values of the primary electric and photometric units in absolute measure, and the legal values for these units shall be those represented by, or derived from, national reference standards maintained by the Department of Commerce.

Then, as a part of the Act of 20 June 1956, the Bureau moved from Washington, DC to Gaithersburg, Maryland. The responsibilities listed in the Act of 21 July 1950, and many others, were transferred to the National Institute of Standards and Technology (NIST) when the National Bureau of Standards was renamed under the guidelines of the Omnibus Trade and Competitiveness Act of 1988:

> The National Institute of Standards and Technology [shall] enhance the competitiveness of American industry while maintaining its traditional function as lead national laboratory for providing the measurement, calibrations, and quality assurance techniques which underpin United States commerce, technological progress, improved product reliability and manufacturing processes, and public safety ... [and it shall] advance, through cooperative efforts among industries, universities, and government laboratories, promising research and development projects, which can be optimized by the private sector for commercial and industrial applications ... [More specifically, NIST is to] prepare, certify, and sell standard reference materials for use in ensuring the accuracy of chemical analyses and measurements of physical and other properties of materials ...

NIST's mission is to promote U.S. economic growth by working with industry to develop and apply technology, measurements, and standards. It carries out this mission through four major programs including ATP, but its centerpiece program is the measurement and standards laboratories program. It provides technical leadership for vital components of the nation's technology infrastructure needed by U.S. industry to continually improve its products and services.

THE ECONOMICS OF STANDARDS [1]

An industry standard is a set of specifications to which all elements of products, processes, formats, or procedures under its jurisdiction must conform. The process of standardization is the pursuit of this conformity, with the objective of increasing the efficiency of economic activity.

---

[1] This section comes directly from the contribution of Gregory Tassey to Audretsch et al. (2002).

The complexity of modern technology, especially its system character, has led to an increase in the number and variety of standards that affect a single industry or market. Standards affect the R&D, production, and market penetration stages of economic activity and therefore have a significant collective effect on innovation, productivity, and market structure. Thus, a concern of government policy is the evolutionary path by which a new technology or, more accurately, certain elements of a new technology become standardized.

Standardization, according to Tassey (2000), can and does occur without formal promulgation as a standard. This distinction between *de facto* and promulgated standards is important more from an institutional process than an economic impact perspective.

In one sense, standardization is a form rather than a type of infrastructure because it represents a codification of an element of an industry's technology or simply information relevant to the conduct of economic activity. And, because the selection of one of several available forms of a technology element as the standard has potentially important economic effect, the process of standardization is important.

While economics is increasingly concerned with standards due to their proliferation and pervasiveness in many new high-technology industries, the economic roles of standards are unfortunately poorly understood. Standards can be grouped into two basic categories:

(1) product-element standards, and
(2) nonproduct-element standards.

This distinction is important because the economic role of each type is different.

Product-element standards typically involve one of the key attributes or elements of a product, as opposed to the entire product. In most cases, market dynamics determine product-element standards. Alternative technologies compete intensely until a dominant version gains sufficient market share to become the single *de facto* standard. Market control by one firm can truncate this competitive process. Conversely, nonproduct-element standards tend to be competitively neutral within the context of an industry. This type of standard can impact an entire industry's efficiency and its overall market penetration rate (Link 1983).

Industry organizations often set nonproduct-element standards using consensus processes. The technical bases (infratechnologies) for these standards have a large public good content. Examples include measurement and test methods, interface standards, and standard reference materials.

From both the positions of a strategically-focused firm as well as a public policy maker, standardization is not an all-or-nothing proposition. In complimented system technologies, such as distributed data processing, telecommunications, or factory automation, standardization typically proceeds in an evolutionary matter in lock step with the evolution of the technology. Complete standardization too early in the technology's life cycle can constrain innovation.

The overall economic value of a standard is determined by its functionality (interaction with other standards at the systems level) and the cost of implementation (compliance costs). Standards should be competitively neutral, which means adaptable to alternative applications of a generic technology over that technology's life cycle.

12 AN INTEGRATED ENTREPRENEURIAL PROCESS

An integrated entrepreneurial process is illustrated in Figure 12.1. As in the earlier schematics in Figure 1.1 and Figure 3.3, the strategic direction of the firm and the competitive pressures that it faces motivate an entrepreneurial response. R&D activity is the primary resource that the firm relies upon to investigate the appropriate response and to act upon it.

Enhancing the firm's R&D activity is its relationship with other organizations and the external environment. One such factor is its involvement in research partnerships, with other firms or perhaps with either a university or a federal laboratory. Firms will also rely on infrastructure technology that comes from federal laboratories, such as NIST (Tassey 1992, 1997), or from the environment created by being located in a science park (Link 1995, 2002). Finally, there is the science base, which consists of the stock of knowledge generated from basic research. The science base resides in the public domain—and the public domain is international in scope—generally in the form of scientific journals but also it is in part embodied in university scientists.

The result of the entrepreneurial process is innovation. An innovation will generate value added if it is accepted in the marketplace, and it will diffuse into society and generate spillover benefits to other firms both within the industry and in outside industries that ultimately use the innovation.

The dashed arrow coming back to the science base shows an internal feedback. Once an innovation exists, knowledge has been created and it too will reside in the public domain.

The usefulness of this illustration is not only as a summary device but also it is a means to highlight the myriad sources of scientific and technical information that firms rely on to support their innovative activity. Certainly, not all firms rely on each source to the same degree. Larger firms in competitive environments generally rely more heavily on their in-house R&D than smaller firms. Small firms rely more on external sources of technical expertise.

Figure 12.1 is also a useful device for summarizing public policies toward R&D. The patent system and the R&E tax credit provide direct incentives to the firm to increase its level of R&D. The NCRA and the NCRPA affect the efficiency of in-house R&D by reducing duplicative research costs and shortening the fundamental research stage. Government-provided infrastructure technology through standards reduces transaction costs in the market, thus lowering the marginal cost of R&D. Finally, federal support of university research continually enriches the science base thus facilitating the in-house R&D process.

Figure 12.1. The Entrepreneurial Process: An Integrated Look

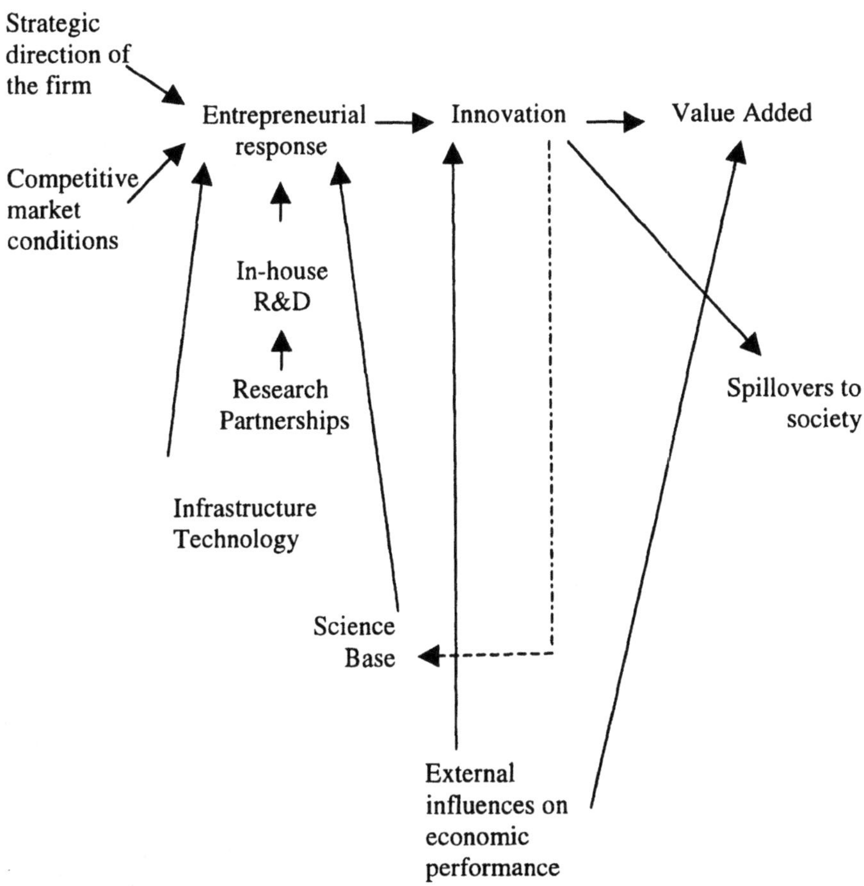

13 LABOR MARKET FOR R&D SCIENTISTS AND ENGINEERS *

*K*ey inputs into the R&D process are scientists and engineers (S&E) working in R&D laboratories. Because of the need for these individuals to be highly trained, one might conclude the relevant segment of that workforce consists of those who are doctoral trained.

There are certain fields in S&E where doctoral training is not a necessary condition for work in R&D. Engineering is a good case in point. Moreover, as innovation becomes more and more imbedded in non-R&D functions of the firm, the doctoral trained workforce may become less key to understanding patterns of innovation.

The majority of doctoral scientists trained in the United States are employed in institutions of higher education, and in business and industry. Over time, the sectoral composition has shifted substantially as industry has employed proportionately more scientists and academe proportionately fewer. While academe once employed 60 percent of all Ph.D. scientists, this percent has been below 50 since 1990. In contrast, industry, which used to employ fewer than one in four scientists, now employs approximately one in three.

The production or supply of new doctorates in the United States can be summarized in terms of the ratio of Ph.D.s granted to U.S. citizens and permanent residents to the U.S. population ages 25 to 34. The proportion of 25 to 34 aged individuals receiving a Ph.D. in both the physical and life sciences increased throughout the 1960s, declined in the 1970s, was fairly stable in the 1980s, and again increased during the 1990s. It is expected to increase in the 2000s.

---

* A portion of this chapter comes directly from the contribution of Paula Stephan to Audretsch et al. (2002).

FOREIGN-BORN SCIENTISTS AND ENGINEERS

Approximately 18 percent of the highly-skilled (doctoral or medical degree) scientists in the United States in 1980 were foreign-born.[1] The percentage of foreign-born was highest among physical scientists (20.4%) and lowest among life scientists (15.4%). By 1990, the proportion foreign-born had increased to nearly 25 percent. More than one in four physical scientists and mathematics and computer scientists working in the United States had been born abroad; for life scientists the proportion had increased from approximately one in seven to one in five. The proportion of engineers who are foreign born is substantially smaller than that of highly-trained (bachelor's degree) scientists. In 1980, approximately 14 percent were foreign-born; this had crept up to about 16 percent by 1990 and may continue to rise.

A large number of immigrants who receive their doctoral training abroad initially come to the United States to take postdoctoral positions. For example, of the 14,918 postdoctoral appointees training in the life sciences in doctorate-training institutions in the United States in 1996, the National Science Foundation (NSF) estimates that 7,425—almost exactly half—were not U.S. citizens. In the physical sciences the proportion was even higher, at 57 percent. While some of these are permanent non-residents, a number come to the United States as temporary residents.

R&D POLICY ON S&E LABOR MARKETS

The federal government has several policy mechanisms to stimulate R&D activity. In industry, research is stimulated through the use of the R&E tax credit as discussed in Chapter 8, as well as through direct subsidies as discussed in Chapter 6 and Chapter 10. University research is encouraged through the provision of a large funding pool for faculty grants. Depending on the sector targeted, the labor market consequences of these policies vary considerably.

---

[1] If they immigrate prior to receiving their doctoral training, and if they indicate at the time that they receive their degree that they plan to remain in the U.S., they are captured in the National Science Foundation's Survey of Doctorate Recipients (the database from where the above statistics came). If, however, they immigrate after receiving their doctoral training, they are not included in this database and are only captured in a sampling frame created every ten years, based on the decennial census and known as the National Survey of College Graduates (NSCG). See Stephan and Levin (2001) for a discussion.

The government's policy intent of stimulating R&D in the private sector, through the policy mechanisms previously discussed, acts to increase the demand for R&D processes and, by extension, the demand for highly-trained scientists and engineers. This increase in demand, however, only leads to an increase in the number of S&E workers in industry *if* the supply of S&E workers is not fixed but instead is responsive to higher wage rates. If the supply of S&E workers is not wage responsive, then public policies to stimulate R&D will only have a compensation effect rather than an employment effect.

Government policy to support R&D in the university sector, however, can affect S&E labor markets quite differently. Graduate students are especially responsive to the level and availability of research support, and a significant portion of the R&D funding that goes to support university research provides support for graduate research assistants.

There are a number of barriers to attracting S&E doctorates into industry (Romer 2000, Stephan and Levin 2000):

- A lack of good information to graduate students concerning the rewards to working in industry.
- Attitudes among faculty that employment in industry makes the student a second-class scientist.
- Opportunities for recent Ph.D.s to remain in academe and work as a post doctorate fellows.
- Funding patterns that tie students and post doctorates to their mentor's laboratory, thereby decreasing the opportunity to have different research experiences and potentially learn more about positions in industry.

The existence of such barriers suggests that government policies as they relate to the S&E workforce are less than efficient. Possible solutions to these inefficiencies include, but are not limited to:[2]

- The provision of information concerning employment opportunities in industry to aspiring students as well as to students enrolled in programs.
- The creation of training grants which fund the student, instead of the faculty member, thereby providing students greater independence.

[2] The variables usually found to affect the supply of enrollees (or the number of graduates) in a specific field are the salary paid in that field, salary in an alternative occupation, such as law or business and (for men) the draft deferment policy.

- The reshaping of graduate education along the lines followed by professional schools, including the provision of information about career prospects and the provision of opportunities to work in industry while in graduate school (Romer 2000).

WOMEN IN SCIENCE

As important as it is for well-trained doctoral students to move into industry, it is equally and in fact more fundamentally important to attract the brightest undergraduate students into science and technology fields. The government is aware of this and has adopted specific programs aimed at, for example, attracting women into doctoral programs in the sciences.

The number of federal programs established to encourage and support women in science has increased in the United States since 1989. For example, the under representation of women in science first gained the national spotlight with the publication of the 1989 White House Report, *Changing America: The New Face of Science and Technology*. Then, in 1991 the National Research Council of the National Academy of Sciences established the Committee on Women in Science and Engineering (CWSE). While not a source of funding, CWSE aggressively encouraged women to pursue science and engineering careers through a series of symposia, workshops, and policy initiatives. Since that time, financial resources available to support graduate studies by women in the sciences have increased.

According to Link and Link (1999), these women in science programs begun in 1989 have had a measurable impact on the number of women completing their doctoral studies.

FORECASTING SCIENTIFIC LABOR MARKETS

Although economic models of scientific labor markets have been somewhat successful in providing insight into factors affecting demand and supply, reliable forecasts of scientific labor markets do not exist, partly because of the unavailability of reliable predictions of variables affecting supply and demand. While this problem is endemic to forecasting in general, the ups and downs of federal funding make forecasts of scientific labor markets particularly unreliable. More importantly, the emergence of new industries, such as bioinformatics and pharmacogenomics, makes it difficult to predict future needs for scientists and engineers and the requisite skill sets.

Despite these concerns, forecasts of scientific labor markets are somewhat common. This occurs, in part, because Congress mandates such forecasts, supposedly in an effort to maintain industrial competitiveness. Also, corporate strategic planning requires periodic estimation of workforce needs.

14 PUBLIC ACCOUNTABILITY

The concept of public accountability can be traced as far back as President Woodrow Wilson's reforms, and in particular to the Budget and Accounting Act of 1921. This Act of June 10, 1921 not only required the President to transmit to Congress a detailed budget on the first day of each regular session, but also it established the General Accounting Office (GAO) to settle and adjust all accounts of the government. We note this fiscal accountability origin because the GAO has had a significant role in the evolution of accountability-related legislation during the past decade.

What follows is a review of the legislative history of initiatives that falls broadly under the rubric of public accountability. As Collins (1997, p. 7) notes:

> As public attention has increasingly focused on improving the performance and accountability of Federal programs, bipartisan efforts in Congress and the White House have produced new legislative mandates for management reform. These laws and the associated Administration and Congressional policies call for a multifaceted approach—including the provision of better financial and performance information for managers, Congress, and the public and the adoption of integrated processes for planning, management, and assessment of results.

Fundamental to any evaluation of a public institution is the recognition that the institution is accountable to the public—that is to taxpayers—for its activities. With regards to technology-based institutions, this accountability refers to being able to document and evaluate research performance using metrics that are meaningful to the institutions' stakeholders, meaning to the public.

PERFORMANCE ACCOUNTABILITY

Chief Financial Officers Act of 1990

The GAO has a long-standing interest and a well documented history of efforts to improve governmental agency management through performance measurement. For example, in February 1985 the GAO issued a report entitled "Managing the Cost of Government—Building An Effective Financial Management Structure" which emphasized the importance of systematically measuring performance as a key area to ensure a well-developed financial management structure.

On November 15, 1990, the 101st Congress passed the Chief Financial Officers Act of 1990. As stated in the legislation as background for this Act:

> The Federal Government is in great need of fundamental reform in financial management requirements and practices as financial management systems are obsolete and inefficient, and do not provide complete, consistent, reliable, and timely information.

The stated purposes of the Act are to:

(1) Bring more effective general and financial management practices to the Federal Government through statutory provisions which would establish in the Office of Management and Budget a Deputy Director for Management, establish an Office of Federal Financial Management headed by a Controller, and designate a Chief Financial Officer in each executive department and in each major executive agency in the Federal Government.

(2) Provide for improvement, in each agency of the Federal Government, of systems of accounting, financial management, and internal controls to assure the issuance of reliable financial information and to deter fraud, waste, and abuse of Government resources.

(3) Provide for the production of complete, reliable, timely, and consistent financial information for use by the executive branch of the Government and the Congress in the financing, management, and evaluation of Federal programs.

The key phrase in these stated purposes is in point (3) above, "evaluation of Federal programs." Toward this end, the Act calls for the establishment of agency Chief Financial Officers, where agency is defined to include each of the Federal Departments. And, the agency Chief Financial Officer shall, among other things, "develop and maintain an integrated agency accounting and financial management system, including financial reporting and internal controls," which, among other things, "provides for the systematic measurement of performance."

While the Act does outline the many fiscal responsibilities of agency Chief Financial Officers, and their associated auditing process, the Act's only clarification of "evaluation of Federal programs" is in the above phrase, "systematic measurement of performance." However, neither a definition of "performance" nor guidance on "systematic measurement" is provided in the Act. Still, these are the seeds for the growth of attention to performance accountability.

Government Performance and Results Act of 1993

Legislative history is clear that the Government Performance and Results Act (GPRA) of 1993 builds upon the February 1985 GAO report and the Chief Financial Officers Act of 1990. The 103rd Congress stated in the August 3, 1993 legislation that it finds, based on over a year of committee study, that:

(1) waste and inefficiency in Federal programs undermine the confidence of the American people in the Government and reduce the Federal Government's ability to address adequately vital public needs;
(2) Federal managers are seriously disadvantaged in their efforts to improve program efficiency and effectiveness, because of insufficient articulation of program goals and inadequate information on program performance; and
(3) congressional policymaking, spending decisions and program oversight are seriously handicapped by insufficient attention to program performance and results.

Accordingly, the purposes of GPRA are to:

(1) improve the confidence of the American people in the capability of the Federal Government, by systematically

> holding Federal agencies accountable for achieving program results;
> (2) initiate program performance reform with a series of pilot projects in setting program goals, measuring program performance against those goals, and reporting publicly on their progress;
> (3) improve Federal program effectiveness and public accountability by promoting a new focus on results, service quality, and customer satisfaction;
> (4) help Federal managers improve service delivery, by requiring that they plan for meeting program objectives and by providing them with information about program results and service quality;
> (5) improve congressional decisionmaking by providing more objective information on achieving statutory objectives, and on the relative effectiveness and efficiency of Federal programs and spending; and
> (6) improve internal management of the Federal Government.

The Act requires that the head of each agency submit to the Director of the Office of Management and Budget (OMB):

> no later than September 30, 1997 ... a strategic plan for program activities. Such plan shall contain ... a description of the program evaluations used in establishing or revising general goals and objectives, with a schedule for future program evaluations.

And, quite appropriately, the Act defines program evaluation to mean:

> an assessment, through objective measurement and systematic analysis, of the manner and extent to which federal programs achieve intended objectives.

In addition, each agency is required to:

> prepare an annual performance plan [beginning with fiscal year 1999] covering each program activity set forth in the budget of such agency. Such plan shall ... establish performance indicators to be used in measuring or

> assessing the relevant outputs, service levels, and outcomes of each program activity;

where "performance indicator means a particular value or characteristic used to measure output or outcome."

Cozzens (1995) speculated, at the time of the passage of GPRA, that it will encourage agencies to ignore what is difficult to measure, no matter how relevant. Alternatively, one could wear a more pessimistic hat and state that GPRA will encourage agencies to emphasize what is easy to measure, no matter how irrelevant.

FISCAL ACCOUNTABILITY

Legislation following GPRA emphasizes fiscal accountability more than performance accountability. While it is not our intent to suggest that performance accountability is more or less important than fiscal accountability, for we believe that both aspects of public accountability are important, our emphasis below is on performance accountability. Nevertheless, our discussion would not be complete in this chapter without references to the Government Management Reform Act of 1994 and the Federal Financial Management Improvement Act of 1996.

Government Management Reform Act of 1994

The Government Management Reform Act of 1994 builds on the Chief Financial Officers Act of 1990. Its purpose is to improve the management of the federal government though reforms to the management of federal human resources and financial management. Motivating the Act is the belief that federal agencies must streamline their operations and must rationalize their resources to better match a growing demand on their services. Government, like the private sector, must adopt modern management methods, utilize meaningful program performance measures, increase workforce incentives without sacrificing accountability, and strengthen the overall delivery of services.

Federal Financial Management Improvement Act of 1996

The Federal Financial Management Improvement Act of 1996 follows from the belief that federal accounting standards have not been implemented uniformly through federal agencies. Accordingly, this Act establishes a uniform accounting reporting system in the federal government.

The above overview of what we call public accountability legislation makes clear that government agencies are becoming more and more accountable for their fiscal and performance actions. And, these agencies are being required to a greater degree than ever before to account for their activities through a process of systematic measurement. For technology-based institutions in particular, internal difficulties are arising as organizations learn about this process.

Compliance with these guidelines is causing increased planning and impact assessment activity and is also stimulating greater attention to methodology. Perhaps there is no greater validation of this observation than the diversity of response being seen among public agencies, in general, and technology-based public institutions, in particular, as they grope toward an understanding of the process of documenting and assessing their public accountability. Activities in recent years have ranged from interagency discussion meetings to a reinvention of the assessment wheel, so to speak, in the National Science and Technology Council's (1996) report, "Assessing Fundamental Science."

SYSTEMATIC APPROACHES TO THE EVALUATION OF TECHNOLOGY-BASED PROGRAMS

GPRA is directionally, as opposed to methodologically, clear about the evaluation process. It stipulates that public institutions/research programs must identify outputs and quantify the economic benefits of the outcomes associated with such outputs. In our opinion, agencies will attempt to quantify outcome benefits and then compare those quantified benefits to the public costs to achieve the benefits. Although these are GPRA's directions, the methodological hurdle that has been plaguing most public agencies is how to quantify benefits. And even with an acceptable quantification of benefits, will the confidence of the American people in public sector research be strengthened by simply comparing benefits to costs?

Consider two different approaches to program evaluation. When evaluating publicly-funded publicly-performed research, the relevant approach is based on a counterfactual method; when evaluating publicly-funded privately-performed research programs, the relevant approach is based on a spillover method. The discussion that follows draws upon Link and Scott (2000).

Traditional Evaluation Methods

Griliches (1958) and Mansfield, et al. (1977) pioneered the application of fundamental economic insight to the development of estimates of private and social rates of return to investments in R&D. Streams of investment outlays through time—the costs—generate streams of economic surplus through time—the benefits. Once identified and measured, these streams of costs and benefits are used to calculate rates of return, benefit-to-cost ratios, or other related metrics.

In the Griliches/Mansfield model, the innovations evaluated are conceptualized as reducing the cost of producing a good sold in a competitive market at constant unit cost. For any period, there is a demand curve for the good and a horizontal supply curve. Innovation lowers the unit cost of production, shifting downward the horizontal supply curve and thereby, at the new lower equilibrium price, resulting in greater consumer surplus (economists' measure of value—the difference between the price consumers would have been willing to pay and the actual price, summed over all purchases, they paid). Additionally, the Griliches/Mansfield model accounts for producer surplus, measured as the difference between the price the producers receive and the actual marginal cost, summed over the output sold, minus any fixed costs. Social benefits are then the streams of new consumer and producer surpluses, while private benefits are the streams of producer surplus, not all of which are necessarily new because the surplus gained by one producer may be cannibalized from the pre-innovation surplus of another producer. Social and private costs will, in general, also be divergent.

The Griliches/Mansfield model for calculating economic social rates of return add the public and the private investments through time to determine social investment costs, and then the stream of new economic surplus generated from those investments is the benefit. Thus, the evaluation question that can be answered from such an evaluation analysis is: What is the social rate of return to the innovation, and how does that compare to the private rate of return? We argue that this is not the most appropriate question to ask from a public accountability perspective. The fact that the social rate of return is greater than the private rate of return may validate the role of government in innovation if the private sector would not have undertaken the research; but it ignores, for example, consideration of the cost effectiveness of the public sector undertaking the research as opposed to the private sector.

The Counterfactual Evaluation Method

A different question should be considered when publicly-funded publicly-performed investments are evaluated. Holding constant the very stream of economic surplus that the Griliches/Mansfield model seeks to measure, and making no attempt to measure that stream, one should ask the counterfactual question: What would the private sector have had to invest to achieve those benefits in the absence of the public sector's investments? The answer to this question gives the benefits of the public's investments—namely, the costs avoided by the private sector. With those benefits—obtained in practice through extensive interviews with administrators, federal research scientists, and those in the private sector who would have to duplicate the research in the absence of public performance—counterfactual rates of return and benefit-to-cost ratios can be calculated to answer the fundamental evaluation question: Are the public investments a more efficient way of generating the technology than private sector investments would have been? The answer to this question is more in line with the public accountability issues implicit in GPRA, and certainly is more in line with the thinking of public sector stakeholders, who may doubt the appropriateness of government's having a role in the innovation process in the first place.

The Spillover Evaluation Method

There are important projects where economic performance can be improved with public funding of privately-performed research. Public funding is needed when socially valuable projects would not be undertaken without it. If the expected private rate of return from a research project falls short of the required rate called the hurdle rate, then the private sector firm will not invest in the project. Nonetheless, if the benefits of the research spill over to consumers and to firms other than the ones investing in the research, the social rate of return may exceed the appropriate hurdle rate. It would then be socially valuable to have the investments made, but since the private investor will not make them, the public sector should. By providing some public funding, thereby reducing the investment amount needed from the private firm or firms doing the research, the expected private rate of return can be increased above the hurdle rate. Thus, because of this subsidy, the private firm is willing to perform the research which is socially desirable because much of its output spills over to other firms and sectors in the economy.

The question asked in the spillover method is one that facilitates an economic understanding of whether the public sector should be underwriting a portion of private-sector firms' research, namely: What proportion of the total profit stream *generated by the private firm's R&D and innovation* does the private firm expect to capture; and hence, what proportion is not appropriated but is instead captured by other firms that imitate the innovation or use knowledge generated by the R&D to produce competing products for the social good? The part of the stream of expected profits captured by the innovator is its private return, while the entire stream is the lower bound on the social rate of return. In essence, our method weighs the private return, estimated through extensive interviews with firms receiving public support regarding their expectations of future patterns of events and future abilities to appropriate R&D-based knowledge, against private investments. The social rate of return weights the social returns against the social investments.

The application of the spillover model to the evaluation of public funding/private performance of research is appropriate since the output of the research is only partially appropriable by the private firm with the rest spilling over to society. The extent of the spillover of such knowledge with public good characteristics determines whether or not the public sector should fund or partially fund the research.

PROGRAM EVALUATION

Many public technology-based institutions have conducted program evaluations both as part of their overall management of the program and in response to GPRA.

The U.S. General Accounting Office (GAO) monitors the performance evaluation progress of federal government agencies. Based on their assessment, "agencies' fiscal year 2000 performance plans show moderate improvements over the fiscal year 1999 plans ... [h]owever, key weaknesses remain ..." (GAO 1999, p. 3). One weakness is that agencies' performance data lack credibility.

Not only is GAO expected to continue to monitor the performance of research programs, but also GPRA-like frameworks are beginning to be used at the state level.

15 CONCLUSIONS

*O*ur purpose in writing this book was to survey the landscape of topics that are related to the field of economics of science and technology. This survey is based on historical and applied perspectives, all within the context of what we refer to as the entrepreneurial process. We hope that this overview will guide students who wish to pursue further research on these topics, and as well will be a source of information for policymakers as they contemplate the bigger picture.

After reading this overview, we also hope that many will reach the same conclusion as we have regarding the state of the field, its emphases and, just as important, the gaps in knowledge. In our judgment, and from the perspective of many researchers in this field, the study of the economics of science and technology has made important advances in the past two decades or so. In some instances, the areas of study we examined herein were essentially unknown as formal research topics just a few short years ago. For example, the study of patents and intellectual property included relatively few important contributions among applied economists until fairly recently. Patent databases and measurement techniques have greatly facilitated the field of study, but equally important has been the influence of public policy. With changes in the laws and statutes pertaining to patents, and particularly the ownership of intellectual property among university and government researchers, the topic has become manifestly more important during recent years. Similarly, the study of domestic technology transfer could barely be considered a field even twenty years ago. But the convergence of research techniques and policy initiative has led to a growth of research on this topic. And most recently, scholars from economics and other disciplines have begun to investigate the firm-university nexus both in terms of technologies that flow from universities into firms via research collaborations thereby leveraging their in-house R&D, as well as in terms of real world knowledge that flows from firms into universities thereby affecting curriculum and the research culture. Technology, or more appropriately, technical knowledge, thus flows two ways; to fully understand the economics of science and technology and its attendant policies both patterns must be recognized and understood.

We believe that there are several issues that require further exploration. The first of these is the relationship between globalization and the development and ownership of science and technology. Although there is a long-standing literature on international technology transfer, the typical model employed in these studies focuses on understanding relations between a technology or knowledge donor nation and a recipient nation. Unfortunately, this type of model is not as relevant or useful for understanding the complexity of the current state of technology transfer as is, because there are now numerous instances where the firms' nation of charter is almost inconsequential, where capital flows in a dizzying array of vehicles, institutions and multinational forums, and where technology is simultaneously marketed along different channels with different firms in a great number of counties. The complexity of this process does not lend itself easily to current models, be they existing neoclassical models in economics or in other disciplines.

Another potentially fruitful area of research for students of the economics of science and technology is e-commerce. Indeed, there is already an increase in interest in electronic commerce and in funding agencies, such as the National Science Foundation, that are girding themselves to support programs of study for this topic. But thus far e-commerce has not been a topic of much interest to more than a handful of researchers using the theories and tools of the economics of science and technology. We suspect that this will change in the next few years.

A third area that requires more attention is the intersection between the economics of science and technology and distributional and social equity issues (beyond the question of how new technology affects the workforce (Siegel (1999)). One of the more interesting questions within the economics of science and technology is who "wins" and who "loses" with innovation and the introduction of new technology. Academics and policy makers have, for example, begun to talk about the "digital divide," but there are also health care technology divides, among others. If we understand the economic forces that allow us efficiently to produce and market health care technologies and pharmaceuticals but do not understand the distributional issues relating to access technology, it is possible to encourage, at the same time, more and more innovation and greater divisions in society.

These are some of the "big questions." Unfortunately, the field is not yet prepared to provide much evidence on these questions. But as needs change, as resources shift, and as more and more informed citizens become aware of the topics we have overviewed in this book, tomorrow's economics of science and technology will almost certainly look very different than today's.

REFERENCES

Abramovitz, Moses. "Resource and Output Trends in the United States since 1870," *American Economic Review*, 1956, pp. 5-23.

Acs, Zoltan, and David B. Audretsch. "Patents as an Measure of Innovative Activity," *Kyklos*, 1989, pp. 171-180.

Acs, Zoltan and David B. Audretsch. *Innovation and Small Firms*, Cambridge: MIT Press, 1990.

Acs, Zoltan, David B. Audretsch, and Maryann P. Feldman. "Real Effects of Academic Research: Comment," *American Economic Review*, 1992, pp. 363-367.

Acs, Zoltan, David B. Audretsch, and Maryann P. Feldman. "R&D Spillovers and Recipient Firm Size," *Review of Economics and Statistics*, 1994, pp. 336-340.

Almeida, Paul and Bruce Kogut. "The Exploitation of Technological Diversity and the Geographical Localization of Innovation," *Small Business Economics*, 1997, pp. 21-31.

Audretsch, David B. *Innovation and Industry Evolution*, Cambridge: MIT Press, 1995.

Audretsch, David B. and Maryann P. Feldman. "R&D Spillovers and the Geography of Innovation and Production," *American Economic Review*, 1996, pp. 630-640.

Audretsch, David B., Albert N. Link, and John T. Scott. "Public/Private Partnerships: Evaluating SBIR-Supported Research," *Research Policy,* forthcoming.

Audretsch, David B. and Paula E. Stephan. "Company-Scientist Locational Links: The Case of Biotechnology," *American Economic Review,* 1996, pp. 641-652.

Audretsch, David B., Barry Bozeman, Kathryn L. Combs, Maryann P. Feldman, Albert N. Link, Donald S. Siegel, Paula E. Stephan, Gregory Tassey, and Charles Wessner. "The Economics of Science and Technology," *Journal of Technology Transfer*, forthcoming 2002.

Austin, David H. "An Event Study Approach to Measuring Innovative Output: The Case of Biotechnology," *American Economic Review Papers and Proceedings*, 1993, pp. 253-258.

Barnett, Homer G. *Innovation: The Basis of Cultural Change*, New York, W.W. Norton, 1953.

Baudeau, Nicolas. *Premier Introduction à la Philosophie Économique*, edited by A. Dubois, Paris: P, Geuthner, 1910 [originally 1767].

Ben-Zion, Uri. "The R&D and Investment Decision and its Relationship to the Firm's Market Value: Some Preliminary Results, in *R&D, Patents, and Productivity*(edited by Z. Griliches), Chicago: University of Chicago Press, 1984, pp. 134-162.

Bound, John, Clint Cummins, Zvi Griliches, Bronwyn Hall, and Adam Jaffe. "Who Does R&D and Who Patents?, in *R&D, Patents, and Productivity* (edited by Z. Griliches), Chicago: University of Chicago Press, 1984, pp. 21-54.

Bozeman, Barry and Albert N. Link. *Investments in Technology: Corporate Strategies and Public Policy Alternatives*, New York: Praeger Publishers, 1983.

Bozeman, Barry and Albert N. Link, "Tax Incentives for R&D: A Critical Evaluation," *Research Policy*, 1984, pp. 21-31.

Brod, Andrew C. and Albert N. Link. "Trends in Cooperative Research Activity: Has the National Cooperative Research Act Been Successful?" in *Innovation Policy in the Knowledge-Based Economy* (edited by M. Feldman and A. Link), Norwell, MA: Kluwer Academic Publishers, 2001, pp. 105-119.

Bush, Vannevar. *Science—the Endless Frontier.* Washington, DC: U.S. Government Printing Office, 1945.

Carr, Robert K. "U.S. Federal Laboratories and Technology Transfer," mimeograph, 1995.

Clark, Burton R. *Places of Inquiry*, Berkeley: University of California Press, 1995.

Cockburn, Iain and Zvi Griliches. "Industry Effects and Appropriability Measures in the Stock Market's Valuation of R&D and Patents," *American Economic Review Papers and Proceedings*, 1988, pp. 419-423.

Cohen, Wesley M. and David A. Levinthal. "Innovation and Learning: The Two Faces of R&D," *Economic Journal*, 1989, pp. 569-596.

Collins, Eileen. "Performance Reporting in Federal Management Reform," National Science Foundation Special Report, mimeographed, 1997.

Council on Competitiveness. *Endless Frontiers, Limited Resources: U.S. R&D Policy for Competitiveness*, Washington, DC: Council on Competitiveness, 1996.

Cozzens, Susan E. "Assessment of Fundamental Science Programs in the Context of the Government Performance and Results Act (GPRA)," Washington, DC: Critical Technologies Institute, 1995.

Crow, Michael and Barry Bozeman. *Limited by Design: R&D Laboratories in the U.S. National Innovation System*, New York: Columbia University Press, 1998.

Dunne, Timothy. "Plant Age and Technology Usage in U.S. Manufacturing Industries," *Rand Journal of Economics*, 1994, pp. 488-499.

Economic Report of the President. Washington, DC: Government Printing Office, 1994.

Economic Report of the President. Washington, DC: Government Printing Office, 1995.

Eden, Lorraine, Edward Levitas, and Richard J. Martinez. "The Production, Transfer and Spillover of Technology: Comparing Large and Small Multinationals as Technology Producers," *Small Business Economics*, 1997, pp. 53-66.

Feldman, Maryann P. "Knowledge Complementarity and Innovation," *Small Business Economics*, 1994, pp. 363-372.

Glaeser, Edward, Heidi Kallal, Jose Scheinkman, and Andrei Shleifer. "Growth of Cities," *Journal of Political Economy*, 1992, pp. 1126-1152.

Grabowski, Henry G. "The Determinants of Industrial Research and Development: A Study of the Chemical, Drug, and Petroleum Industries," *Journal of Political Economy*, 1968, pp. 156-159.

Griliches, Zvi. "Research Costs and Social Returns: Hybrid Corn and Related Innovations," *Journal of Political Economy*, 1958, pp. 501-522.

Griliches, Zvi. "Issues in Assessing the Contribution of R&D to Productivity Growth," *Bell Journal of Economics*, 1979, pp. 92-116.

Griliches, Zvi. "Market Value, R&D, and Patents," *Economics Letters*, 1981, pp. 183-187.

Griliches, Zvi. "Productivity, R&D, and Basic Research at the Firm Level in the 1970s," *American Economic Review*, 1986, , pp. 141-154.

Griliches, Zvi. "Patent Statistics as Economic Indicators: A Survey," *Journal of Economic Literature*, 1990, pp. 1661-1707.

Hagedoorn, John, Albert N. Link, and Nicholas S. Vonortas. "Research Partnerships," *Research Policy*, 2000, pp. 567-586.

Hall, Bronwyn H., Zvi Griliches, and Jerry A. Hausman. "Patents and R&D: Is There a Lag?" *International Economic Review*, 1986, pp. 265-302.

Hall, Bronwyn H., Adam Jaffe and Manuel Trajtenberg. "Market Value and Patent Citations: A First Look" National Bureau of Economic Research Working Paper No. 7741, June 2000.

Hall, Bronwyn H., Albert N. Link, and John T. Scott. "Universities as Research Partners," NBER Working Paper No. 7643, April 2000.

Hall, Bronwyn H., Albert N. Link, and John T. Scott, "Barriers Inhibiting Industry from Partnering with Universities: Evidence from the Advanced Technology Program," *Journal of Technology Transfer*, 2001, pp. 87-98.

Hall, Bronwyn H. and John van Reenen. "How Effective are Fiscal Incentives for R&D? A Review of the Evidence," *Research Policy*, 2000, pp. 449-469.

Hébert, Robert F. and Albert N. Link. *The Entrepreneur: Mainstream Views & Radical Critiques*, second edition, New York: Praeger Publishers, 1988.

Henderson, Rebecca, Adam Jaffe and Manuel Trajtenberg, "Universities as a Source of Commercial Technology: A Detailed Analysis of University Patenting, 1965-1988," *Review of Economics and Statistics*, 1998, pp. 119-127.

Hounshell, David A. "The Evolution of Industrial Research in the United States," in *Engines of Innovation: U.S. Industrial Research at the End of an Era* (edited by R. Rosenbloom and W. Spencer), Boston: Harvard Business School Press, 1996, pp. 13-86.

House Committee on Science. *Unlocking Our Future: Toward a New National Science Policy*, Report to Congress, September 24, 1998.

Jaffe, Adam B. "Technological Opportunity and Spillovers of R&D: Evidence from Firm's Patents, Profits, and Market Value," *American Economic Review*, 1986, pp. 984-1001.

Jaffe, Adam. "Real Effects of Academic Research," *American Economic Review*, 1989, pp. 957-970.

Jaffe, Adam B., Michael S. Fogarty, and Bruce A. Banks. "Evidence From Patents and Patent Citations on the Impact of NASA and other Federal Labs on Commercial Innovation," *Journal of Industrial Economics*, 1998, pp. 183-206.

Jaffe, Adam, Manuel Trajtenberg, and Rebecca Henderson. "Geographical Localization of Knowledge Spillovers as Evidenced by Patent Citations," *Quarterly Journal of Economics*, 1993, pp. 577-598.

Kaufer, Erich. *The Economics of the Patent System*, Chur, UK, Harwood Academic Publishers, 1989.

Kerr, Clark. *Troubled Times in American Higher Education*, Albany: State University of New York Press, 1994.

Leyden, Dennis P. and Albert N. Link. "Tax Policies Affecting R&D: An International Comparison," *Technovation*, 1993, pp. 17-25.

Leyden, Dennis P. and Albert N. Link. "Federal Laboratories as Research Partnerships," *International Journal of Industrial Organization*, 1999, pp. 575-592.

Lichtenberg, Frank R. and Donald Siegel. "The Impact of R&D Investment of Productivity-New Evidence Using Linked R&D-LRD Data," *Economic Inquiry*, 1991, pp. 203-228.

Link, Albert N. "Basic Research and Productivity Increase in Manufacturing," *American Economic Review*, 1981, pp. 1111-1112.

Link, Albert N. "Market Structure and Voluntary Product Standards," *Applied Economics*, 1983, pp. 393-401.

Link, Albert N. *A Generosity of Spirit: The Early History of Research Triangle Park*, Research Triangle Park, NC: Research Triangle Foundation of North Carolina, 1995.

Link, Albert N. *Evaluating Public Sector Research and Development*, New York, Praeger Publishers, 1996a.

Link, Albert N. "On the Classification of Industrial R&D," *Research Policy*, 1996b, pp. 397-401.

Link, Albert N. "Public/Private Partnerships in the United States," *Industry and Innovation*, 1999, pp. 191-217.

Link, Albert N. *From Seed to Harvest: The Growth of Research Triangle Park,* Research Triangle Park, NC: Research Triangle Foundation of North Carolina, forthcoming 2002.

Link, Albert N. and Laura L. Bauer. *Cooperative Research in U.S. Manufacturing*, Lexington, MA: Lexington Books, 1989.

Link, Albert N. and Jamie R. Link. "Women in Science: An Exploratory Analysis of Trends in the United States," *Science and Public Policy,* 1999, pp.

Link, Albert N. and John Rees. "Firm Size, University Based Research and the Returns to R&D," *Small Business Economics*, 1990, pp. 25-32.

Link, Albert N. and John T. Scott. *Public Accountability: Evaluating Technology-Based Institutions*, Norwell, MA: Kluwer Academic Publishers, 1998.

Link, Albert N. and John T. Scott. "Evaluating Federal Research Programs," mimeographed, 2000.

Link, Albert N. and John T. Scott. "Public/Private Partnerships: Stimulating Competition in a Dynamic Economy," *International Journal of Industrial Organization*, 2001, pp. 763-794.

Link, Albert N. and Gregory Tassey. *Strategies for Technology-based Competition: Meeting the New Global Challenge*, Lexington, MA: Lexington Books, 1987.

Mansfield, Edwin. *Industrial Research and Technological Change*, New York: W.W., Norton, 1968.

Mansfield, Edwin. *Technological Change*, New York: W.W. Norton, 1971.

Mansfield, Edwin. "Basic Research and Productivity Increase in Manufacturing," *American Economic Review*, 1980, pp. 863-873.

Mansfield, Edwin, John Rapoport, Anthony Romeo, Samuel Wagner, and George Beardsley. "Social and Private Rates of Return from Industrial Innovations," *Quarterly Journal of Economics*, 1977, pp. 221-240.

Morgan, Robert P. "University Research Contributions to Industry: The Faculty View," in *Trends in Industrial Innovation: Industry Perspectives & Policy Implications* (edited by P. Blain and R. Frosch), Research Triangle Park: Sigma Xi, The Scientific Research Society, 1998, pp. 163-170.

Morrison, Catherine and Donald Siegel. "External Capital Factors and Increasing Returns in U.S. Manufacturing," *Review of Economics and Statistics*, 1997, pp. 647-654.

Mowery, David C. and David J. Teece. "Strategic Alliances and Industrial Research," in *Engines of Innovation: U.S. Industrial Research at the End of an Era* (edited by R. Rosenbloom and W. Spencer), Boston: Harvard Business School Press, 1996, pp. 111-129.

Mueller, Dennis C. "The Firm Decision Process: An Econometric Investigation," *Journal of Political Economy*, 1967, pp. 58-87.

National Resources Committee. *Research—A National Resource.* Washington, DC: Science Committee of the National Resources Committee, November 1938.

National Science and Technology Council. "Assessing Fundamental Science," Washington, DC: National Science and Technology Council, 1996.

National Science Board. *Science & Engineering Indicators—2000*, Washington, DC: National Science Foundation, 2000.

National Science Foundation. *Graduate Studies and Postdoctorates in Science and Engineering: 1996*, NSF 98-301. Washington, DC: National Science Foundation, 1998.

Nelson, Richard R. *National Innovation Systems: A Comparative Analysis*, Oxford: Oxford University Press, 1993.

Nightingale, Paul. "A Cognitive Model of Innovation," *Research Policy*, 1998, pp. 689-709.

Office of the President, *U.S. Technology Policy*, Washington, DC: Executive Office of the President, September, 26, 1990.

Office of Technology Policy. "Effective Partnering: A Report to Congress on Federal Technology Partnerships, Washington, DC: U.S. Department of Commerce, 1996.

Olson, Kristen. "Total Science and Engineering Graduate Enrollment Falls for Fourth Consecutive Year," *Data Brief*, Arlington: National Science Foundation, December 17, 1998.

Pakes, Ariel. "On Patents, R&D, and the Stock Market Rate of Return," *Journal of Political Economy*, 93, 1985, pp. 390-409.

Pakes, Ariel. "Patents as Options: Some Estimates of the Value of Holding European Patent Stocks," *Econometrica*, 1986, pp. 755-784.

Pakes, Ariel and Zvi Griliches. "Patents and R&D at the Firm Level: A First Report," *Economics Letters*, 1980 (5), pp. 377-381.

Pakes, Ariel and Zvi Griliches. "Patents and R&D at the Firm Level: A First Look," in *R&D, Patents, and Productivity* (edited by Z. Griliches), Chicago: University of Chicago Press, 1984, pp. 55-72.

Pakes, Ariel and Mark Schankerman. "The Rate of Obsolescence of Patents, Research Gestation Lags, and the Private Rate of Return to Research Outcomes," *R&D, Patents, and Productivity* (edited by Z. Griliches), Chicago: University of Chicago Press, pp. 73-88.

Prevezer, Martha. "The Dynamics of Industrial Clustering in Biotechnology," *Small Business Economics*, 1997, pp. 255-271.

Reis, Paula and D.H. Thurgood. *Summary Report 1991: Doctoral Recipients from United States Universities*, Washington DC: National Academy Press.

Romeo, Anthony. "Interindustry and Interfirm Differences in the Rate of Diffusion of an Innovation," *Review of Economics and Statistics*, 1975, pp. 311-319.

Romer, Paul M. "Should the Government Subsidize Supply or Demand in the Market of Scientists and Engineers?" NBER Working Paper No. 7723, May 2000.

Saxenian, Annalle. "Regional Networks and the Resurgence of Silicon Valley," *California Management Review*, 1990, pp. 89-111.

Schankerman, Mark and Ariel Pakes. "The Rate of Obsolescence and the Distribution of Patent Values: Some Evidence from European Patent Renewals," *Revue Economique*, 1985, pp. 917-941.

Schankerman, Mark and Ariel Pakes. "Estimates of the Value of Patent Rights in European Countries During the Post-1950 Period," *Economic Journal*, 1986, 96, pp. 1052-1076.

Scherer, F. M. "Firm Size, Market Structure, Opportunity and the Output of Patented Inventions," *American Economic Review*, 1965, pp. 1097-1125.

Scherer, F. M. "Inter-Industry Technology Flows and Productivity Growth," *Review of Economics and Statistics*, 1982, pp. 627-634.

Scherer, F. M. "The Propensity to Patent," *International Journal of Industrial Organization*, 1983, pp. 107-128.

Schmookler, Jacob. *Invention and Economic Growth*, Cambridge, MA: Harvard University Press, 1966.

Schwalbach, J. and Klaus F. Zimmermann. "A Poisson Model of Patenting and Firm Structure in Germany," in *Innovation and Technological Change: An International Comparison* (edited by Z. Acs and D. B. Audretsch), Ann Arbor: University of Michigan Press, 1991, pp. 109-120.

Scott, John T. "Financing and Leveraging Public/Private Partnerships: The Hurdle-Lowering Auction," *STI Review*, 1998, pp. 67-84.

Siegel, Donald S. "The Impact of Computers on Manufacturing Productivity Growth: A Multiple-Indicators, Multiple-Causes Approach," *Review of Economics and Statistics*, 1997, pp. 68-78.

Siegel, Donald S. *Skill-Biased Technological Change: Evidence from A Firm-Level Survey*, Kalamazoo, MI: W.E. Upjohn Institute Press, 1999.

Siegel, Donald S. and Zvi Griliches. "Purchased Services, Outsourcing, Computers, and Productivity in Manufacturing," in *Output Measurement in the Service Sector,* Chicago: University of Chicago Press, 1992, pp. 429-458.

Siegel, Donald S., Jerry G. Thursby, Marie C. Thursby, and Arvids A. Ziedonis. "Organizational Issues in University-Industry Technology Transfer: An Overview of the Symposium Issue," *Journal of Technology Transfer*, 2001, pp. 5-11.

Siegel, Donald S., David A. Waldman, and Albert N. Link, "Assessing the Impact of Organizational Practices on the Productivity of University Technology Transfer Offices: An Exploratory Study," NBER Working Paper No. 7256, July 1999. (Revised version forthcoming in *Research Policy*.)

Sobel, Dava. *Longitude*. New York: Penguin Books, 1995.

Solow, Robert M. "Technical Change and the Aggregate Production Function," *Review of Economics and Statistics*, 1957, pp. 312-320.

Steelman, John R. *Science and Public Policy: A Program for the Nation.* Washington, DC: The U.S. Presidents Scientific Research Board, August 27, 1947.

Stephan, Paula E. "The Economics of Science," *Journal of Economic Literature*, 1996, pp. 1199-1235.

Stephan, Paula E. and Sharon G. Levin. "Exceptional Contributions to U.S. Science by the Foreign-Born and Foreign-Educated," *Population Policy and Research Review*, 2001, pp. 59-79.

Tassey, Gregory. *Technology Infrastructure and Competitive Position*, Norwell, MA: Kluwer Academic Publishers, 1992.

Tassey, Gregory. *The Economics of R&D Policy*, Westport, CT: Quorum Books, 1997.

Tassey, Gregory. "R&D Trends in the U.S. Economy: Strategies and Policy Implications," Gaithersburg, MD: National Institute of Standards and Technology, 1999.

Tassey, Gregory. "Standardization in Technology-Based Markets," *Research Policy*, 2000, pp. 587-602.

Terleckyj, Nestor E. *Effects of R&D on the Productivity Growth of Industries: An Explanatory Study*, Washington, DC: National Planning Association, 1974.

Tibbetts, Ronald, "The Small Business Innovation Research Program and NSF SBIR Commercialization Results," mimeograph, 1999.

Trajtenberg, Manuel. "A Penny for Your Quotes: Patent Citations and the Value of Innovations," *Rand Journal of Economics,* 1990a, pp. 172-187.

Trajtenberg, Manuel. *Economic Analysis of Product Innovation: The Case of CT Scanners,* Cambridge, MA: Harvard University Press, 1990b.

Unesco. *National Science Policies of the U.S.A.: Origins, Development and Present Status*. Paris: Unesco, 1968.

U.S. Congress, Office of Technology Assessment. *The Effectiveness of Research and Experimentation Tax Credits*, Washington, DC: Office of Technology Assessment, 1996.

U.S. Department of Commerce. "Emerging Technologies: A Survey of Technical and Economic Opportunities," Washington, DC: Technology Administration, 1990.

U.S. General Accounting Office (GAO). *Managing for Results: Opportunities for Continued Improvements in Agencies' Performance Plans*, Washington, DC: General Accounting Office, 1999.

Wessner, Charles W. *The Small Business Innovation Research Program: An Assessment of the Department of Defense Fast Track*, Washington, DC: National Academy Press, 2000.

White House Task Force on Women, Minorities, and the Handicapped in Science and Technology. *Changing America: The New Face of Science and Technology*, Washington, DC: The Task Force, 1989.

INDEX

MIX
Papier aus verantwortungsvollen Quellen
Paper from responsible sources
FSC® C105338

If you have any concerns about our products,
you can contact us on
ProductSafety@springernature.com

In case Publisher is established outside the EU,
the EU authorized representative is:
Springer Nature Customer Service Center GmbH
Europaplatz 3, 69115 Heidelberg, Germany

Printed by Libri Plureos GmbH
in Hamburg, Germany